# THE NATURE OF
# WAR

Conflicting Paradigms and
Israeli Military Effectiveness

# THE NATURE OF WAR

Conflicting Paradigms and
Israeli Military Effectiveness

RON TIRA

Copyright © The Institute for National Security Studies, 2010, 2014.

The right of Ron Tira to be identified as Author of this work has been asserted in accordance with the Copyright, Designs and Patents Act 1988.

2 4 6 8 10 9 7 5 3 1

*First published in hardcover 2010, reprinted in paperback, 2014 in Great Britain by*
SUSSEX ACADEMIC PRESS
PO Box 139
Eastbourne BN24 9BP

*Distributed in North America by*
SUSSEX ACADEMIC PRESS
ISBS Publisher Services
920 NE 58th Ave #300, Portland, OR 97213, USA

All rights reserved. Except for the quotation of short passages for the purposes of criticism and review, no part of this publication may be reproduced, stored in a retrieval system, or transmitted, in any form or by any means, electronic, mechanical, photocopying, recording or otherwise, without the prior permission of the publisher.

*British Library Cataloguing in Publication Data*
A CIP catalogue record for this book is available from the British Library.

*Library of Congress Cataloging-in-Publication Data*
Tira, Ron.
  The nature of war : conflicting paradigms and Israeli military effectiveness / Ron Tira.
    p. cm.
  Includes bibliographical references and index.
  ISBN 978-1-84519-638-7 (pbk. : alk. paper)
   1. Military art and science—Israel. 2. Military doctrine—Israel.
 3. Israel—Military policy. 4. Israel—Armed Forces. 5. Israel—Defenses.
 6. Israel—History, Military. 7. Politics and war—Israel. I. Title.
 UA853.I8T57 2010
 355'.03355964—dc22

2009025782

Typeset and designed by Sussex Academic Press, Brighton & Eastbourne.

# Contents

*Acknowledgments* — vi

Executive Summary — 1

Introduction: The Erosion of Classical Military Doctrine — 11

1  Doctrinal Background — 15

2  "Simple" Symmetrical Wars — 31

3  The Complex Asymmetrical War against a Regular Opponent: The Picture becomes Multidimensional — 37

4  Asymmetrical Wars against Non-State Opponents: Same Theater of Operations, Different Objectives — 69

5  Parallel War: One War with Two Non-Convergent Campaigns — 77

6  The Second Lebanon War and Operation Cast Lead: Parallel Wars against a Non-State Opponent — 85

7  The Future War: Parallel War against a State Enemy that has Adjusted to Fighting against RMA and Adopted a Guerilla Paradigm — 109

Conclusion: It Never Ends — 123

*Notes* — 132
*Index* — 138

# Acknowledgments

I would like to extend my heartfelt thanks to Maj. Gen. (ret.) Giora Eiland. This memorandum owes its existence to his guidance and support. Likewise, I would like to thank Brig. Gen. (ret.) Dr. Dov Tamari, Maj. Gen. (ret.) Giora Romm, Maj. Gen. (ret.) Haim Erez, Maj. Gen. (ret.) Aharon Ze'evi Farkash, Col. (ret.) Shlomo Kashi, Prof. Azar Gat, Dr. Anat Kurz, and Lt. Col. Roni Amir, the Campaign Planning Department's chief of doctrine (Air Force) for their time, comments, and insights.

# THE NATURE OF
# WAR

# Executive Summary

Classical Clausewitz doctrine of war paints an orderly, universal, causal, and somewhat simple picture: victory in war means the fulfillment of the war's political objective, and it is achieved through military decision. Decision means a significant enough blow to the enemy's military to render it incapable of operating effectively against us. Such a decision is achieved by attacking the enemy's operational center of gravity, usually in the prime major battle. The two armed forces align themselves symmetrically in the same theater of operations with the same goal in mind: mutual attack against the military centers of gravity, in order to render a blow to the opposing force's field effectiveness.

This doctrine has proved itself in some instances, though not in all. The purpose of this monograph is to analyze the cases and the circumstances where more complex analytical and managerial tools are necessary to deal with these cases, taking into consideration Israel's particular circumstances and those of any other nation under discussion. The monograph begins with a survey of a few types of war and the circumstances whereby classical doctrine is progressively less valid, and then attempts to devise additional analytical tools necessary to understand these same circumstances. On the basis of these new tools and concepts, the study then examines the relevance of classical doctrine to two central cases. The first is the Second Lebanon War (2006), where the accepted analytical tools failed to reflect the context and complexity of the situation, and where Israel viewed the war as not much more than a list of targets that needed to be attacked, without dealing with the need to undermine the enemy's war paradigm. This is followed by a brief analysis of Operation Cast Lead in Gaza, testing whether Israel exhibited a more methodological strategic and operational approach than in the Second Lebanon War. The second posited case is the next generation of wars that Israel and other Western countries may find themselves fighting – wars against states that have adopted the guerilla paradigm (reminiscent of Hizbollah's approach).

*Some of the major points raised by this monograph are*:

- War does not necessarily test only military effectiveness on the battlefield, but often other issues as well. These issues may be a test of the ability to endure and outlast an extended war of attrition,[1] or a test of the ability to harness the international community to forge the political end state, or a test of the scope and recruitment of war-enabling resources of the battling sides, and so on.

The main military force application thread helps direct the war towards the selected test. This thread may be, for example, attacking the enemy's forces directly or indirectly, or denying the enemy's strategic freedom of action to continue fighting; attempting to influence the desire for war of the enemy's political echelon, or attempting to undermine public support for continuing the war by raising its price; attacking war-enabling resources and international supply lines, and so on. The issue to be tested and the main force application thread always depend on context and are a function of the relative strengths and weaknesses of the particular sides. Military decision is usually relevant in wars testing the military effectiveness of the sides, while in other types of war no decision is achieved quickly enough, so that other elements (such as stamina, resources, recruitment of the international community) determine the outcome of the war. The center of gravity under attack in a war is of course affected by the issue being tested and by the force application thread.

- In the background of every war is a struggle over the nature of the war and the issues it is going to test. War must be channeled towards the nature and issues where one enjoys a relative advantage.

In World War II, Germany sought to test the sides' military effectiveness, but in the end the war hinged on the scope and pace of recruiting national resources and on national stamina. In the Vietnam War, the United States sought to test the sides' military effectiveness and national resources, but the war ultimately hinged on the sides' civilian–political stamina. In the Yom Kippur War, Israel also sought to test the sides' military effectiveness, but Egypt succeeded in navigating the war in such a way as to hinge on the sides' stamina and their ability to recruit the international community to shape the political end state. Thus, the first imperative is to understand and successfully shape the nature of the war.

- Impacting negatively on the enemy's battlefield fitness is not necessarily enough to achieve decision and victory. War is a multi-level phenomenon, and success at one level does not ensure success at another.

War has many levels: the grand strategy level, the military strategy level, the operational level, the tactical level, the techno-tactical level, and other context-dependent levels (logistics, information, and so on). The classical approach assumes that amassing successes at one level translates into success at other levels. Thus, the traditional assumption is that scoring success at the tactical level translates into success at the operational level; scoring operational success translates into strategic success; and enough strategic successes translate into the opportunity to realize the political objective of the war. However, despite the obvious reciprocal relationships between the various levels of war, the levels are also somewhat isolated from one another. Sometimes, there is a struggle at each of the levels with its own autonomous rationale specific to that level, and it is difficult to translate success at one level into an achievement at another level. For example, in the Yom Kippur War, the southern front confrontation ended with results that were essentially different from one another at the various levels: Israel achieved a military decision at the operational level, while Egypt realized its plans and objectives at the military strategy and grand strategy levels, and won the war. Therefore, it is necessary to deal with and overcome the enemy at each level separately, and according to the specific playing field at every single level.

- In classical warfare, it was necessary to wait for a clear military outcome and only then, and as a result, was it possible to shape the political end state. In a multilevel war, it is sometimes possible to shape the political end state before the military outcome is clear, and even with no relation to it.

In wars with loose connections between the levels or in wars that do not test military effectiveness only, a situation may arise where under certain circumstances the political end state does not derive from the military end state, and the military moves are meant to serve only as the context and catalyst. Again, the Yom Kippur War on the southern front demonstrates this, as does Security Council Resolution 1701, which professed to regulate the strategic–political end state of the Second Lebanon War and did not emerge directly from the outcomes of the battles.

- The question of symmetry/asymmetry entails more than just two situations (a war between two state militaries or between a state and a non-state enemy), but is characterized by many intermediate circumstances. As the asymmetry grows the relevance of classical doctrine declines, even in a war between regular armed forces.

The asymmetry in warfare between regular state armed forces may begin with an asymmetry in the rationale for the war (Germany–Soviet Union) or in the nature of the war (Germany–France, 1940), and reach its climax in "parallel warfare," a situation in which each side manages a separate campaign, different in features and fought parallel to one another, to attack different centers of gravity of the enemy, without the two campaigns necessarily converging in the same space and undermining one another.

## EXECUTIVE SUMMARY

Alternatively, the forces do converge in the same theater of operations but realize different objectives at asymmetrical levels from the encounter. The classic example of parallel warfare is the Second Punic War (218–201 BCE), when Hannibal conducted a campaign against Rome in Italy while Scipio Africanus conducted a campaign against Hannibal in the Iberian Peninsula and North Africa, posing a direct threat to the capital, Carthage. In contrast, campaigns were conducted between states and non-state enemies with profoundly classical symmetry. Examples are the First Lebanon War and the Dien Bien Phu campaign (1953–54).

- Warfare becomes complex especially in conflicts between Western militaries and Asian and Middle Eastern cultures, which view conflicts from a wider perspective than simply a clash between military formations in a theater of operations limited in time and space.

Western countries have not fought one another in the last sixty years, and even industrialized countries (Western or not) have hardly battled against each other in the last few decades. Unlike the West, Asian and Middle Eastern cultures, not raised in the Clausewitz tradition, view war as a comprehensive clash between civilizations that intentionally draws in civilians on both sides, blurs the line between warfare and diplomacy and between war and peace, and is neither limited to the battlefield nor delimited by clear dates. Mao Tse-tung claimed that the view that strategic victory is determined by tactical success is mistaken and ignores the fact that victory is first and foremost a question of whether the "situation as a whole" has been adequately taken into consideration.

- The technological, techno-tactical, and tactical superiority of the West and Israel propels potential enemies to shape wars that avoid a test of the effectiveness of military fielded formations, and rather test primarily both sides' stamina and the ability to outlast each other on the civilian and political dimensions.

The enemy suffering from military inferiority but enjoying greater stamina tries to create a situation that on the one hand is exhausting and politically unbearable for long, and on the other hand denies us the opportunity for military decision and a military exit. In order to avoid the test of decision, the enemy does not show up for the prime major battles, endeavors to conceal and blur the operational centers of gravity, and strives not to operate as a system that may be disrupted. At times, the enemy is capable of attaining the war's objectives even without achieving a military decision in its favor. It does so by the very fact that it does not capitulate for a long enough time – so that parameters such as national stamina or the intervention of the international community determine the outcome of the war.

# EXECUTIVE SUMMARY

This is the basis for the phenomenon of "winning by not losing." The Vietnam War, the Yom Kippur War on the southern front, and the Second Lebanon War demonstrate this principle.

- Both the American doctrine and the classical Israeli approach try to avoid wars that test the stamina of the political–civilian systems, i.e., wars of mutual attrition. However, the circumstances creating attrition are different in the case of a superpower and in the case of a small country.

The United States conducts one-sided wars beyond its own shores (i.e., against enemies that usually do not have the strategic attack capability that can harm the US itself). Therefore, from an American perspective, the very appearance on the ground at the theater of operations is what enables the enemy's potential for reciprocal blows – and attrition; hence the preferred alternative is a one-sided strategic attack on the enemy's leadership and war resources, taking advantage of the staying power afforded by superpower status. In contrast, the meaning of Israel's geo-strategy (a lack of depth) is that even the enemy's tactical capabilities are liable to create an opposing strategic threat. Unlike the United States, even when Israel makes techno-tactical use of standoff fire, at the operational and strategic levels, Israel will never be able to conduct its wars on a standoff basis and will always be exposed to reciprocal strikes. Therefore, Israel must rapidly eliminate the enemy's threat, including by means of confrontation with the enemy's fielded force. Without depriving the enemy of its military capability or at least the window of opportunity to strike at Israel, a reciprocal strategic attack will eventually be created, one that tests both nations' stamina, and this, from Israel's perspective, is the meaning of attrition. Although for a superpower a test of the sides' staying power lies at the core of its relative advantage, for Israel this is a highly undesirable challenge.

- For Israel, almost every type of warfare that tests something other than military effectiveness constitutes an undesirable war of attrition.

These "other" wars are characterized by the enemy's shying away from the test of military decision. The war continues long enough until other factors, such as stamina, resources, or the leveraging of the international community determine the war's outcome – and in all of these, Israel is usually at a relative disadvantage. The core of Israel's defense doctrine must therefore be the effort to direct the war towards testing military effectiveness. Israel's war plans must focus on attacking the enemy's paradigm and denying the enemy the freedom of action and the opportunity to wage the war of attrition it seeks. In order to erode the Israeli civilian and political

## EXECUTIVE SUMMARY

home front, the enemy must prolong the war; Israel, however, must bring about its quick decision. Thus, the time factor becomes a central element in shaping the nature of the war and the issues tested. This fact challenges Israel's preference for using standoff fire from its territory, which over the years became a popular mode of warfare, as campaigns based on firepower alone tend to be protracted and ultimately test the endurance of both sides.

- The Clausewitz characterization of decision in terms of destruction of tactical mass or in physical terms is not sufficient.

Classical doctrine defines decision as an essential blow to the enemy's capability of acting effectively against us, typically referring to the enemy's center of mass. The Von Naumann tables for estimating the collapse of military units, in common use since the nineteenth century, claim that eliminating 60 percent of the force of a military unit is equivalent to its destruction. This is an example of the tactical-physical dimension, which in some cases is no longer useful for assessing decision. Particularly in those wars where the enemy does not attempt to defeat Israel's military in the theater of operations but rather seeks to attack its civilian–political stamina, the enemy is not required to maintain full military fitness, and as long as the enemy retains residual defiance capability, it may well operate effectively to realize its plans. Thus, for example, it is possible to question whether eliminating 60 percent of Hizbollah's rocket capability, which during the Second Lebanon War would have reduced the number of launches at the Israeli home front from 250 to 100 per day, would have meant a victory against Hizbollah. It would not seem so, and the operational and strategic benefit that Hizbollah would have gained from either 250 or 100 launches per day would have been quite similar. In other words, even 100 rockets a day would have left Hizbollah with residual defiance capability, enabling the organization to operate with strategic effectiveness to realize its objectives. Therefore, the challenge is how to administer a blow to the enemy's capability to operate effectively while the enemy continues to achieve the required strategic benefit, even when its tactical or operational effectiveness is dealt a significant blow. This reality also renders investment in defensive measures (such as rocket interception) less effective at the strategic level. Even a successful interception system cannot ensure hermetic defense: the penetration of "only" dozens of rockets would be enough for the enemy to achieve the desired effect (civilians going into bomb shelters, being evacuated, the disruption of normal civilian life) and could challenge a state's civilian–political staying power.

- The concept of military decision remains relevant primarily as an analytical tool for clarifying the components that will make it possible to deny the enemy the capability to realize its plans at the higher levels of war.

The effectiveness of military force is not generic or universal, but depends on context, the nature of the war, the issue being tested, and the prevailing paradigm. Therefore, the meaning of the term decision is bound also to denying the enemy the capability or opportunity of realizing the paradigm of the war and the objective it seeks.

- In classical doctrine, the center of gravity was usually the center of the enemy fielded force's mass or a physical or functional Achilles' heel in its operational formation. Today, however, even the state enemy tends to conceal operational centers of gravity and develop amoeba-like qualities (i.e., transition to a network of autonomous cells while lacking a systemic backbone). Therefore, the concept of a center of gravity must also serve as an analytical tool for identifying the point whose attack would deny the enemy the strategic freedom of action to continue fighting, or would at least undermine the enemy's paradigm of war.

The phenomena described above curtail the expected benefit and expected achievement at the operational and strategic levels from attacking the enemy's forward fielded echelon. This reality forces a search for alternate centers of gravity to attack, including physical strategic centers of gravity and non-physical centers of gravity, such as the enemy's plans and basic assumptions of war. These centers of gravity are difficult to identify and cannot be contained in a standardized check list. They are context-dependent and must be analyzed before and during the course of each war.

- Not every asset, as valuable as it may be, is a center of gravity. In a war whose nature lies in testing the military effectiveness of the sides, the center of gravity is defined by the fact that its being attacked or threatened has a direct and immediate effect on the military progress of the war (i.e., on the military capabilities of the enemy or its strategic freedom of action).

By contrast, an attack on assets, which affects primarily the cost of the war or aspires to affect the political drive of the enemy, is more a question of the endurance and staying power of the sides in a protracted war of attrition rather than a component of quick decision. Scipio's threat against Hannibal's capital, Carthage, lay in its immediate and physical conquest, so that Hannibal could not ignore the threat and was forced to abandon his Italian campaign. In contrast, the American aerial attacks on Hanoi during the Vietnam War raised the price of the war but did not represent an immediate and physical threat forcing North Vietnam to stop fighting. Rather, it left the North Vietnamese leadership the option of paying the price and continuing to fight. Indeed, it is difficult to undermine the will of a third world dictatorship to fight once it has made the decision to go war and is willing to pay the price.

- Israel failed in the Second Lebanon War in part because it viewed the war as not much more than a list of targets to attack with standoff fire. The failure lay not only in the lack of effectiveness of standoff fire in the particular circumstances of that war, but primarily in viewing the war through the narrow prism of a target list and nothing else.

Therefore, a return to ground mobile warfare (maneuver[2]) is indeed essential, and maneuver is an irreplaceable tool in the military tool chest for determining the nature of the war. However, if next time Israel only aims to attack that same target list with low trajectory short range fire instead of standoff fire, then it has not learned enough. It is necessary to assess the war in all its complexity and on all its levels, and not only as a list of targets to attack – whether with standoff fire or with low trajectory fire. Thus, for example, in the Second Lebanon War, Israel inadvertently accepted Hizbollah's attempt to dictate the nature of the war as one that tested, through the exchange of fire, the stamina and staying power of the sides over time, a test that Israel did not really want. Even if the Israel Defense Forces (IDF) had executed a successful limited maneuver in southern Lebanon, it would still have been operating within Hizbollah's plans rather than attacking those plans. Israel did not succeed in undermining the basic assumptions and defeating Hizbollah's paradigm. And, Israel did not succeed in imposing a different type of war on the organization, one that was better suited for Israel.

- The next generation of war that Israel and other Western countries are liable to find themselves fighting are wars against a regular state military that has adopted the guerilla paradigm. Such a war challenges the basic assumptions of the classical doctrine with challenges such as those described above, and necessitates more complex planning and analytical tools.

A direct physical blow to such an enemy's capability to act towards realizing its objectives at the higher levels of war seems difficult or even impossible, and therefore decision and victory are located elsewhere – in denying the enemy the strategic freedom of action to fight, in upsetting the enemy's war paradigm and imposing a different type of war, and in attacking centers of gravity different from those known to us from the

# EXECUTIVE SUMMARY

"simple" wars of the past. Ironically, it is precisely the adoption of the guerilla model by the state enemy that makes it difficult to arrive at restrained decision against its fielded echelon, and is liable to turn the war into something more extreme.

Figure 0.1 charts the interface between different levels of war and central relevant issues that determine the war's outcome.

|  | Military effectiveness | National resources | Stamina | Enlisting the international community |
|---|---|---|---|---|
| The grand strategy level |  | World War II in its entirety ↓ | The Vietnam War | ←Kosovo |
| The military strategy level | The Second Punic War |  |  | The Yom Kippur War (Israel–Egypt) ↑ |
| The operational level | The German–French campaign, 1940 | ←The campaign in the Atlantic, 1940-43 | ←The German–Soviet campaign |  |
| The tactical level | The Six Day War |  |  |  |

**Figure 0.1** The War Matrix: The issue and level where the outcome was determined

# Introduction
## The Erosion of Classical Military Doctrine

Clausewitzian military theory[3] tries to create a universal, structured, and causal order for planning and waging wars on the basis of a sequence of key basic assumptions:

1. Victory in war means realizing the political objectives of the war.
2. Victory in war is achieved through military decision.
3. Military decision means a critical blow to the enemy force's capability of operating against us effectively.
4. Military decision is achieved through attacking the enemy's military centers of gravity, usually located in the war's theater of operations.
5. The most common method for attacking the enemy's military center of gravity is through the prime major battle.

This study argues that the validity of these basic assumptions, referred to as classical military doctrine, has eroded. The monograph tests whether there is in fact a direct causal link between these assumptions. Also discussed are different aspects to characterizing military decision and characterizing the enemy's center of gravity that should be attacked, and to what extent the method of attacking the center of gravity is indeed the prime major battle. Two further questions deepen the understanding of classical military doctrine: does attacking the enemy's center of gravity require both sides to present themselves in the same theater of operations? and when both sides are present in the same theater of operations, are they there for the same objective?

This study also examines political aspects of war, but only to the extent necessary to understand its military side.[4] One of the primary tests of war is the ability to leverage and harness the international community to support and facilitate the strategic–political end state of the war. This is a complex test, consisting of different elements, among them the ability to persuade foreign governments of the existence of common interests and the ability to build a fighting or diplomatic coalition (the intergovernmental axis), as well as the ability to enlist public opinion, direct the information flow to the public from the battlefield and shape the war narrative, and lay claim to the legitimacy of the war (the government-public axis). These issues may well be fluid and temporary, and often the ability to take advantage of them depends on context, timing, and extent

## INTRODUCTION

of the variables involved. Thus, for example, the legitimacy that Israel enjoyed during the Second Lebanon War and has enjoyed since the withdrawal from the Gaza Strip was not enough to enable it to achieve its objectives and emerge victorious. Another primary test is that of national endurance, analyzed briefly in this memorandum. Stamina too depends on context and is variable and fluid: in World War I, France displayed remarkable stamina, while only two decades later, in World War II, the civilian and military leadership of France and even the French public capitulated within a few weeks.

Classical military doctrine is disputed, and the debate has sharpened in relation to at least four pivotal developments: (a) increased firepower and obstacles, as well as battlefield saturation of forces in World War I; (b) the outbreak of the nuclear age; (c) the proliferation of conflicts between states and non-state and irregular enemies; (d) over the last fifteen years, with the development of the Revolution in Military Affairs (RMA[5]) in its various incarnations and with its varied components. The debate touches not only on the causal connection between the basic assumptions enumerated above, but also on the content of each of the assumptions and even on the terminology used. It is impossible in a memorandum of this scope to survey all the differences of opinion as to those basic assumptions, and only a few are mentioned.

In order to examine classical doctrine's ability to meet complex contemporary tests, the study below reviews several important interim points. Chapter 1 surveys portions of the commonly accepted doctrinal and theoretical background. Chapter 2 examines "simple" and symmetrical wars in which the assumptions of classical doctrine largely prevailed. Chapter 3 examines a number of wars between states that showed signs of deviating from classical doctrine; these wars were characterized primarily by a loosening of the connections between the developments at the different levels of war, and by the fact that the wars tested broader issues than just the field effectiveness of the fighting sides. Chapter 4 examines war between states and non-state enemies, particularly guerilla forces. Chapter 5 describes the indirect approach and the parallel war, i.e., a war in which each side conducts a separate and parallel campaign against the enemy, without the two campaigns converging or directly interfering with one another. After these surveys and an examination of the concepts and analytical tools that stem from them, the memorandum reaches two main points: in Chapter 6, the Second Lebanon War tests classical doctrine, while in Chapter 7, the memorandum presents a thesis concerning the nature of the next generation of wars and their challenges – wars against regular state enemies that have adopted the guerilla paradigm – and tests the validity of the Clausewitz doctrine to those cases. The study's thesis is also tested against Israel's recent experience in Gaza during Operation Cast Lead.

The monograph claims that a doctrine of warfare cannot be universal because national ways of war and the issues each nation seeks to test in its wars vary and are dependent on circumstances and on relative strengths and weaknesses. If so, a theory of war cannot exist in an empty abstract space and is thus inseparable from the national-subjective context.

This study does not compare symmetrical and asymmetrical wars per se. Nonetheless, the issue of asymmetry (in its more expanded definition presented below) is a convenient tool for organizing the examples surveyed here and is also relevant for dealing with the issues discussed below. Finally, while making references to theoretical and academic work in this field, this study does not purport to represent a comprehensive in-depth academic survey, but to serve as a practical aid for campaign planners and those in similar functions.

# 1

# Doctrinal Background

This chapter explores how the classical Clausewitz approach and more modern approaches relate to the terms victory, decision, center of gravity, the prime major battle, and the other basic assumptions enumerated in the Introduction, above. It describes how Israel's traditional defense concept relates to these issues, as well as certain relevant aspects of American doctrine, the concept of RMA as reflected in the strategic attack doctrine of the US air force, and systemic operational design or operational art.

In addition, the chapter discusses the claim that a doctrine cannot be abstract and universal, rather depends on circumstance and the relative strengths and weaknesses of every nation involved. Issues covered here are the vast gulf separating the United States and Israel in context of the types of war; the term "decision"; what constitutes a "war of attrition"; which centers of gravity should be attacked; and which patterns of action should be followed to avoid attrition.

## Clausewitz and the American Doctrine

Clausewitzian doctrine defines war as a violent way to achieve political objectives. It assumes that the enemy's primary center of gravity is the military mass concentrated in the operational space (the theater of operations) of the war, thus making the core of the war the meeting between the opposing militaries in the same theater of operations and for the same objective, namely: to attack the central mass of the enemy's force.

Indeed, Clausewitz claims that the principal way of achieving victory in war is through destroying the enemy's military:[6] "We concluded that the grand objective of all military action is to overthrow the enemy – which means destroying his armed forces."[7] However, according to Clausewitz, destroying the enemy's military does not mean the complete annihilation of its fighting force up to the last soldier, rather crushing the force's original structure so that it can no longer execute its designated objective. An examination of Clausewitz's prescription for crushing the enemy military's structure reveals that he primarily has in mind those enemy formations we are liable to meet in battle in the war's theater of operations. Thus, for example, he determines that "a theater of war, be it large or small, and the forces stationed there, no matter what their size, represent the sort of unity in which a single center of gravity can be identified. That is the place where the decision should be reached."[8] In addition, the center of gravity is often the point where the enemy mass is the most concentrated, and consequently, "the major battle is therefore to be regarded as concentrated war, as the center of gravity of the entire conflict."[9] While Clausewitz's body of work describes a more complex concept, this description shall suffice for our current purpose.

As defined by American doctrine,[10] the purpose of the military force in war is to prevail over the enemy force, to destroy its fighting capability, and to conquer or retain territory in order to change the enemy's government or its policies. The military end state of every campaign is defined as the point at which the president no longer needs military tools to realize national objectives (this definition differs from the definition of "decision"). The military commander infers the military end state necessary to bring about the national end state from the strategic national end state defined by the president. The strategic national end state and the end state in military operational level terms are achieved by a campaign that consists of a series of major operations all put into action using the same rationale.

The American doctrine described here defines the center of gravity as the source of power providing freedom of action, physical force, and the will to fight. The loss of the center of gravity leads to defeat. The doctrine distinguishes between strategic and operational centers of gravity. The strategic center of gravity is likely to be the military, the allies in the fighting coalition, the leadership, critical capabilities, or the national will. The operational center of gravity may, in addition, be a critical component of the enemy's array of forces. A center of gravity is discernible by the capabilities it imparts, namely, in that it enables the enemy to resist an end state, and by its vulnerability, i.e., in that it is open to attack. The American doctrine distinguishes between a physical center of gravity, such as the concentration of a mass of enemy armed forces or the capital city, and an abstract center of gravity, such as the will to fight. The operational approach is the preferred method for dealing with centers of gravity.

Thus, the center of gravity concept is an analytical tool meant to assist in analyzing the enemy's loci of strengths and weaknesses against which the systemic or operational design as well as major operations should be directed. In other words, the process of analysis is two staged: first, it is necessary to identify the enemy's center of gravity that must be attacked; second, it is necessary to identify the operational method that will enable attacking the enemy's center of gravity effectively.

The American doctrine makes use of the new term "decisive point" and defines it as that place, event, factor, or function that enables the attainment of a critical advantage or contributes to the realization of the objectives (this definition also differs from the definition of "decision"). Decisive points are not centers of gravity, but attacking them is an indirect way of arriving at the center of gravity when it is not possible to do so directly. A "line of operation" is the characterization of force application, particularly via major operations, in a way that passes through the decisive points and achieves the military end state.

Despite the centrality of "decision" to the classical doctrine described above, American documents on doctrine generally do not define and do not even use the term – and later it will be clear why – and therefore we must look for the definition of decision elsewhere. Kober, who reviews the doctrinal background of the term decision, suggests the following definition: "Military decision in a war is denying the enemy's fighting capacity during the war on the battlefield by military means, so that its recovery in the course of the same war is highly unlikely."[11] Kober defines the center of gravity under attack in order to achieve decision as "a weakness that if attacked or captured . . . upsets and disrupts the entire structure of enemy deployment, to the point of denying the enemy the capability of continuing to fight."[12] The term "fighting capability," which Kober does not define explicitly, is usually defined as a combination of ability to fight and the will to do so. Nonetheless, in classical documents, the term "will" usually means the will of the enemy forces in the field that are directly exposed to the menace of war, and not the political will of the enemy's decision makers. The focus on the will of the enemy's leadership is usually a direct consequence of much later thinking than that characterizing Clausewitz's time.

## The Israeli and German Approach

The classical way of thinking was adopted in Israel long ago. Tal, for example, asserts that

> [Israel's] doctrine of defense obligates us to seek quick military decisions with the aim of concluding wars in their early stages, while inflicting painful defeats on the enemy and eliminating significant parts of enemy forces and conquering parts of enemy territory. . . . Eliminating a military force means destroying its organization and denying it the capability to function according to its designated purpose. Destroying a force . . . removes the direct and immediate threat to our existence, and therefore this is clearly an important and "useful" military objective.[13]

The classical approach is also reflected in the statement made in the early 1980s by the IDF chief of the Planning Directorate: "No strategy and no operational or tactical move of one kind or another will prevent the frontal encounter between us and our enemy . . . No indirect approach either, successful as it may be, can avert the major frontal encounter in which the ground forces are designated to play a central role – the role of decision."[14]

While a large number of American doctrine documents fall short of defining "decision" explicitly (though the meaning may sometimes be inferred through a definition of objectives), decision does appear as a basic term in the doctrines of smaller countries, such as Israel and Prussia/Germany of the 19th and early 20th centuries. The reason for this stems in part from the difference in circumstances. A superpower fighting weaker enemies is capable under certain circumstances of realizing its war objectives even without needing to achieve a military decision against enemy forces in their entirety. Thus, for example, in the 1991 Gulf War, the United States achieved the objective of the war (the liberation of Kuwait) by defeating the Iraqi forces stationed in the theater of operations (Kuwait and the adjacent area to its north), without needing to achieve a military decision against the Iraqi military as a whole. A superpower is also capable of fighting several enemies simultaneously and defeating them all.

However, with Israel and Prussia/Germany the approach differs. Despite the obvious differences in size between Israel and Germany, the two countries perceived themselves as small, with few resources and limited stamina, surrounded by several fronts where larger enemies with better staying power were operating simultaneously. These circumstances led Israel and Prussia/Germany to adopt a military strategy that strives to conduct multi-front wars sequentially, by moving the main effort from front to front during the course of the war. The basic condition for thinning out one's forces on a particular front and moving the effort away from it is a blow to the enemy's capability to continue being an effective threat on that particular front, or in other words, achieving decision against the enemy military in its entirety. Only when the enemy force is completely defeated on one front is it possible to turn one's back on it and focus on the next front. This is how decision against the enemy force was promoted to the rank of a basic principle of military doctrine. Moreover, given the circumstances described, the concept of decision also touches on shaping the nature of the war at its highest level – the attempt to create a series of short, quick tests of the military effectiveness of the warring sides while avoiding tests of the belligerents' strengths in staying power, scope of their national resources, stamina, and so on.

Indeed, the core of Prussian/German military strategy at the end of the nineteenth century and the beginning of the twentieth century was moving its main force between the fronts and achieving rapid sequential decisions in large battles of maximum force versus maximum force. This was intended to limit the length of the war as a whole and to achieve unmistakable successes in a way that would curb the appetite of other potential enemies to join the circle of war. (Moltke called the strategy of moving forces and concentrating them consecutively on different fronts the "system of expedients."[15])

For Israel, turning decision into a basic principle of doctrine was the result of three additional rationales. First, Israel strove to expand the time lapse between wars, and therefore took advantage of the opportunity afforded by war to achieve as extensive destruction of the enemy force as possible in order to extend the time needed for its rehabilitation. Second, in the 1950s, when its defense concept was formulated, Israel was exhausted by protracted low intensity fighting (against the fedayeen), and sought to exchange a protracted and exhausting conflict testing its stamina for a short, high intensity war in which decision would be achieved against the Egyptian military. Third, Israel's grand strategy assumed that amassing a series of unequivocal military successes would achieve "grand deterrence" that in turn would steer Arab grand strategy away from the confrontational route towards the diplomatic route (what finally happened with Egypt after an Israeli decision in four wars). Thus it would have been possible, for example, to open the Straits of Tiran for shipping even with an operation limited to the shores of the Gulf of Eilat, but Israel inclined towards opening the shipping channel by means that also included decision against the Egyptian military in its entirety. The desire for decision against the enemy force in its entirety thus became a principle with a life of its own that was nearly as important as realizing the specific objectives of the impending war.

## Operational Art and RMA

The ideas of Shimon Naveh and others regarding the development of advanced military thinking, or operational art, also return in the end to some classical basic assumptions.

> The idea of the systemic blow defines in practical terms a consequent state of a fighting system that is no longer capable of achieving its objectives. This result . . . comes about through a process in which the operational level maneuver represents the dominant operational element . . . The depth includes the space in which the system's fighting mass is deployed. This is the source of the notion "center of mass" . . . Further, depth also defines the theater of operations in which the operational level maneuver . . . will take place . . . The notion of center of gravity consists of [key] elements, [including]: a precise identification of the enemy system's strengths and weaknesses . . . [and] taking advantage of those weaknesses by inflicting maneuvering blows . . . A systemic weakness means identifying a particular situation created by an accumulation of certain systemic

## DOCTRINAL BACKGROUND

circumstances that encourages dealing a blow that will cancel the defeated system's ability to carry out its original mission.[16]

Although this excerpt makes use of the term "depth," it is clear from the context that what is referred to is the depth of the theater of operations (i.e., the operational space of a war where the enemy force is concentrated), and that "depth" in this context differs from "strategic depth," which is used below.

Another contemporary attempt to deviate from classical military doctrine is the American concept of RMA in its various incarnations and components. RMA is based on the technological leap that has been made in the capabilities of guided precision strike weaponry, originally meant to destroy the echelons deployed in the depth of the Soviet operational system and intended to be committed to battle at a later stage. However, with the end of the Cold War, the capabilities of attacking from a distance were applied in a much more far reaching manner. RMA became an alternate way of looking at war, striving to attack the enemy's strategic centers of gravity directly as well as undermine the enemy's functioning and rationale as a system without actually taking up a position and first maneuvering around the battlefield against the enemy force. Not surprisingly, this way of thinking primarily characterizes elements of long-range fire, and particularly aerial fire. And indeed, even though RMA has come into its own only in the last decade and a half, it is possible to discern the raw form of this thinking about conducting a campaign of strategic fire against the state home front (rather than against the fielded forces of the enemy military) as early as among the first theoreticians of aerial warfare in World War I, such as Giulio Douhet of Italy, Hugh Trenchard of Britain, and Billy Mitchell of the US.

In the spirit of RMA, the strategic attack doctrine of the US Air Force (Document AFDD 2-1.2, 2007)[17] states that strategic attacks directly aimed against strategic centers of gravity generate effects at the strategic level and achieve, or directly support the achievement of, objectives at the strategic level, without the need for achieving operational objectives as an interim stage. Strategic attack becomes possible thanks to the unique capabilities of aerial forces to realize objectives by striking at the heart of the enemy, by disrupting the vital activity of its leadership, its war supporting resources, and its strategy, and by bypassing the need to fight through levels of the enemy's fielded formations. According to this doctrine, a strategic center of gravity is defined as a point that holds the enemy's system or structure together, and supplies the purpose and direction for the entire enemy system. The doctrine of strategic attack emphasizes the centrality of attacking enemy war-supporting resources as an indirect way of affecting its political will.

AFDD 2-1.2 clarifies that a direct confrontation with the enemy's fielded force exposes one's own vital targets to the enemy's counteroffensive, and requires the allocation of more resources to the fighting. The document contends that if the classic force-on-force conception of war entailed mutual attrition until the realization of the objectives of the war, then strategic attack enables the circumvention of a confrontation with the enemy's military force and strikes against other elements of enemy strength instead.

However, perhaps the most interesting light shed on RMA was by an earlier, 2003 version of the AFDD 2-1.2, which stated:

> While other uses of military power ultimately seek to attain national objectives, they do so through the accumulation of tactical and/or operational-level effects against enemy military forces. Strategic attack, in contrast, seeks to achieve conflict objectives without focusing on attrition of or direct engagement with enemy military forces. Strategic objectives are thus achieved "most directly" – without the traditional necessity of defeating the enemy in force-on-force conflict . . . Until the advent of airpower, strategic effects in wartime could only be achieved by battling through enemy forces to the centers of gravity they protected, or by subduing the enemy through exhaustion or attrition. Strategic attack offers the option of bypassing this more traditional form of warfare and striking directly at the heart of the enemy in some conflicts and affords a vital complement to more traditional means in others."[18]

Similarly:

> Military forces exist to protect centers of gravity, preserve freedom of action, or enable strategy. In many cases, it is necessary to first disable at least some portion of enemy fielded forces through tactical action before being able to directly pursue strategic objectives. . . . Conflicts based on Clausewitz's dictum are focused upon engaging enemy forces. Defeat of these forces becomes an end unto itself, and the *ends* of the conflict may thus be lost in the effort to defeat the *means*. Strategic attack seeks more direct achievement of the ends – by defeat of the enemy's will through mechanisms other than engaging the enemy's fielded forces.[19]

## The American Doctrine and the Israeli Experience

However, there is an essential difference in the circumstances and perspectives of attrition between the United States and Israel, and consequently, an essential difference regarding the prescribed main force application thread and the preferred center of gravity for attack. The two nations both strive to avoid protracted tests of the political–civilian system's stamina,

## DOCTRINAL BACKGROUND

but the circumstances bringing about such a test would be very different. The United States usually conducts its wars across the ocean, against smaller enemies lacking the capability of taking direct action against the United States itself. From the American perspective, the very fact of sending expeditionary forces into a tactical battlefield enables the enemy to act reciprocally and effectively against the United States and strike a blow at its forces, i.e., enables the enemy to achieve attrition. The American experience proves that inflicting harm on ground forces is liable to harm civilian and political support for continuing the war and potentially cause the Americans to withdraw from the challenge. Therefore, the preferred American alternative is direct and one-sided attack from the air on the enemy's leadership and war-enabling resources, without risking the lives of Americans. In such a way, the capability of conducting the war by standoff methods at all war's levels is attained. The Serbs, for example, lacked the ability to return effective fire not only against the attacking planes but also, and in particular, against the bases in Italy from which the planes took off, or against Washington or New York. Without pictures of funerals and flag-draped coffins aboard transport planes, there is less pressure on – and by – the American public, and the United States enjoys almost unlimited staying power. The one-sided strategic attack using the American steamroller of fire can last weeks and even months under the diplomatic, economic, and logistical umbrella of a superpower's stamina. When a superpower attacks a small nation one-sidedly, the channeling of the war so as to test the sides' resources, their stamina, and their ability to enlist the international community in effect means channeling the war towards the core of the superpower's relative strength.

In contrast, the geo-strategic reality of Israel (and the lack of strategic depth in particular) turns even tactical capabilities the enemy has into a strategic threat and grants the enemy freedom of action against Israel itself. Therefore, Israel must strike a blow at the enemy's military operating capability, or at least narrow the window of opportunity available to the enemy to act against Israel, even by taking up a position on the tactical battlefield for direct or indirect action against the enemy forces. Without striking a blow against the military operating capability of the enemy, there is a risk of mutual strategic attack ultimately testing the national stamina of the two warring sides. From the Israeli perspective, this is the undesirable war of attrition, as Israel has limited staying power – diplomatically, logistically, financially, and also in terms of the vulnerability of the home front. It is precisely the American alternative for attrition (attacking the will of the enemy leaders and the enemy's war-supporting resources) that in Israel's circumstances creates a war hinging on the issue of both warring sides' respective staying power. In other words, from Israel's perspective, this constitutes a war of attrition.

Moreover, the American doctrine of strategic attack lends a great deal of weight to attacking enemy strategic war resources, while in the background there is a model reminiscent of the circumstances of World War II (described below). However, this approach has very limited applicability to Israel's wars. In the Middle Eastern reality, the local war industry does not affect the war at a time constant relevant to the course of the confrontation (which lasts at high intensity for no more than a few weeks); most of the resources needed to continue the fighting are already in the hands of the fielded units as the war opens, and their dependence on the supply of additional resources from the strategic level is limited. Moreover, in the new Arab paradigm of war based on avoidance of major battles and on prolonged defiance, the dependence on resources is limited in any case.

In the 1970s and 1980s, the intensive defense dialogue between the United States and Israel focused on the Americans studying the Israeli lessons of war. Israel shared its experience of struggling with the Arabs who were reliant on Soviet doctrines and weapons. Israel also shared its experience with situations that were applicable to the United States. For example, the 1973 campaign against Syria resembled to some extent what NATO could expect in case of a surprise Warsaw Pact attack on West Germany. Starting in the 1990s, the trend reversed itself, and Israel began to learn the American lessons of war, with American RMA concepts starting to trickle into Israeli thinking about defense. However, the American experience is problematic as a case study applicable to Israel, for a variety of reasons.

The Iraqi military in both the 1991 and the 2003 wars was not able to cope with new capabilities of intelligence gathering, firepower, and RMA, and did not try to present a war paradigm that would undermine the American relative advantage and attack the United States' war plans. On the contrary, it "volunteered" to present a relatively passive and massive deployment with a high signature on an exposed plain, and to play the part the Americans had written for it. The Iraqi military took up positions in the theaters of the 1990s and 2000s against the best equipped and most organized fighting force in the world, with a war paradigm suited for equal peers in the 1970s. Moreover, the scope of national resources enlisted by the United States was completely disproportionate in comparison with the resources available to Iraq.

Neither are the lessons of Kosovo (Operation Allied Force, 1999) particularly relevant to Israel. It is possible that what best clarifies the nature of force application used in Kosovo and its rationale was the disagreement between the Supreme Allied Commander, General Wesley Clark, and the commander of the Joint Allied Air Forces, General Michael Short, about the preferred mode of action for the airpower.[20] Clark wanted to deal a direct blow against the Serbian forces' capabilities in Kosovo to continue with the program of ethnic cleansing, while Short wanted to deal a blow to the Serbian leadership's will to continue with the ethnic cleansing,

## DOCTRINAL BACKGROUND

by using protracted punitive strikes in Belgrade and against national Serbian assets. In the end, the airpower was used primarily according to General Short's view, both as a matter of preference and because of the lack of effectiveness that became apparent by the attempt to use airpower alone against the Serbian fielded forces. To this day, it is not clear why Milosevic finally capitulated. Perhaps it was a result of the 38,000 sorties carried out by NATO, perhaps it was the threat of a ground invasion, or perhaps it was due to the Russian withdrawal of support and Serbia's international isolation. In any case, the war did not hinge on a test of the military effectiveness of the warring sides, because the air campaign was not at all effective against the Serbian forces in Kosovo, which continued their program of ethnic cleansing essentially with almost no interference during the entire period of the fighting. The capability of the Serbian forces to act effectively in fulfilling their objectives was not impaired, or in doctrinal terms, no military decision was reached against them. Instead, in a completely one-sided manner, over 78 days the United States tested the Serbian regime's stamina and staying power against American firepower, and NATO's ability to enlist the international community, in particular Russia, to support the desired end state was also tested.

This model of war is not suited to Israel, because in contrast to the American use of one-sided fire in Kosovo, any firepower-based campaign that Israel would be a party to would be bilateral and reciprocal. Unlike the United States, even when Israel makes techno-tactical use of standoff fire, at the strategic and operational levels, Israel can never conduct its wars by standoff methods and will always be exposed to counterstrikes. This leads to three conclusions: first, unlike the United States in Kosovo, Israel cannot relinquish the idea of dealing a blow to the enemy force's capability of operating effectively against it (namely, a direct or indirect decision against the enemy force); second, directing the war towards a mutual protracted test of the sides' stamina and staying power and a test of their ability to enlist the international community means directing the war towards challenges in which Israel usually suffers from a glaring disadvantage and its enemies usually have a clear relative advantage; and third, as to the time factor: exchanges of fire tend to be drawn out over long periods of time, but Israel does not have the resources and stamina for the 44 months of aerial bombardment afforded (unsuccessfully) by Operation Rolling Thunder in the Vietnam War, or even for the 78 days of Kosovo. Such conclusions must be part of the foundation of Israel's defense concept.

The story of the 1991 and 2003 wars with Iraq and the war in Kosovo is therefore the story of a sole superpower, enjoying absolute supremacy and a large edge as regards national resources, stamina, and ability to enlist the international community. It is the story of a superpower with an ocean or two separating it from its theaters of operations, which fights at the strategic level one-sidedly by means of a fire-reach of global proportions and expeditionary forces. It is possible that Israel inadvertently adopted defense concepts rooted in the American experience. Israel did not study only the techno-tactics, but apparently also the American way of war. These American constructs never suited Israel's circumstances, and were in fact among the factors responsible for the failures of the Second Lebanon War (2006). Indeed, as a result of that war, many essays have been written critical of Israel's adoption of elements of RMA, including by this author,[21] and therefore a critical examination of RMA is not at the center of this memorandum and RMA will be mentioned only in later sections. Nonetheless, it is useful to devote some attention to an examination of what is shared by classical military doctrine, operational art, and RMA.

## Common to the Different Approaches

First, classical military doctrine, operational art, and RMA all view the planning of the war as a two-stage analytical process whereby the enemy's center or centers of gravity, which if attacked enables realization of the war's objectives, must be defined. Then, the operational method that enables striking the said centers of gravity in the most effective way possible must be characterized. All the approaches share this process of thinking that hinges on the enemy's centers of gravity, even if each doctrine defines different centers of gravity according to its particular world view.

Second, despite the essential differences in theoretical backgrounds and terminologies, none of the approaches strives to achieve total annihilation of the enemy force to its very last fighter, rather to create a situation in which the enemy force stops functioning in a manner that prevents us from realizing our objectives or loses its capability to fulfill its objectives. In Clausewitz's terms, this entails crushing the original structure of the enemy's forces to render it incapable of fulfilling its designated mission. To use the terms surveyed by Kober, this entails depriving the enemy of its fighting capability. According to Tal, annihilating the fighting force means destroying its organization and depriving it of the capability to function according to its defined purpose. To Shimon Naveh, it entails elimination of the defeated system's capability to carry out its original mission. And in terms of AFDD 2-1.2, it entails locating vulnerability nodes in the enemy system and attacking them in order to suppress its functional effectiveness and denying the enemy its systemic rationale.

One can certainly claim that these definitions are essentially similar. Even though Clausewitz tended to view the enemy military's mass as its center of gravity, it is clear that in certain cases the gap narrows between the Clausewitz search for a center of gravity and the "critical/vulnerability nodes" of RMA's Effect Based Operations (EBO),[22] so that the essential difference between the enemy's loss of military capability to fulfill its missions, according to Clausewitz, and denying the enemy its systemic rationale, according to EBO, is eroded. The terms "the enemy's plans," "the enemy's paradigm," and "the enemy's systemic rationale" in fact almost overlap, and the terms "center of gravity" and "the constituents of decision" simply provide perspective on these terms. We are dealing with rich and varied terminology ultimately describing similar elements. A brief example will illustrate this claim.

One of the Allies' principles for landing on the beaches of Normandy in Operation Overlord (1944) was isolating the beachhead and denying the German mobile reserves (most of whom were situated on the far side of the Seine river) the opportunity to attack it. This was done by directly attacking these reserves from the air, by a prior attack on rail lines, tunnels, bridges, and other possible critical transit points between the reserves and the beachhead, and by parachuting forces deep behind the German front-line isolating the landing beaches. Thus the Allies succeeded in creating a situation in which the landing forces had to deal primarily with the German front line units. The German commander on the Western front, Field Marshal Gerd von Rundstedt and his mobile reserves were denied the opportunity to fulfill their designated purpose as a system and as a front and to act effectively against the beachhead. This thread of operation of the Allies can be described similarly using classical doctrine terms (attacking the enemy's operational center of gravity so that the enemy's military cannot fulfill its function), operational art terms (operational level maneuver into the depth of the enemy system, denying it the capability of fulfilling its mission), and also in Effect Based Operations terms (as a blow to vulnerability nodes in order to defeat the enemy system rationale). This is thus a case when the terms center of gravity and decision (meaning an essential blow to the enemy force's capability of acting against us effectively to fulfill its objective) are a certain ramification of the same, congruent doctrinal definitions.

Parenthetically, it seems almost as if Clausewitz already tried to deal with different and more complex approaches than his own (such as RMA[23]) and even tried to reach common ground as early as the beginning of the nineteenth century, writing:

> How are we to counter the highly sophisticated theory that supposes it possible for a particularly ingenious method of inflicting minor direct damage on the enemy's forces to lead to major indirect destruction; or that claims to produce, by means of limited but skillfully applied blows, such paralysis of the enemy's forces and control of his will-power as to constitute a significant shortcut to victory? Admittedly, an engagement at one point may be worth more than at another. Admittedly, there is a skillful ordering of priority of engagements in strategy; indeed, that is what

strategy is all about, and we do not wish to deny it. We do claim, however, that direct annihilation of the enemy's forces must always be the dominant consideration. We simply want to establish this dominance of the destructive principle.[24]

## So What Do We Do in War?

While this chapter's discussion might seem to the reader to be theoretical and of limited practical application in the real world, it is not, and the questions raised in it are critical to the process of making the most significant decisions in the shaping and planning of war. The following example demonstrates the point.[25]

In 1904–6, France and Britain stood on the brink of war against Germany because of a crisis in Morocco. Concerned that the Moroccan kingdom was on the verge of dissolution, Britain and France sought to protect their vital interests, especially those connected to the Straits of Gibraltar and the ports of North Africa. However, Germany started to intervene in developments in Morocco, threatening the interests of the two colonial powers. Fundamentally, they were thus facing a much larger issue than Morocco: Britain and France sought to curb the ambitions of the new European power – recently united under Bismarck – to become a significant player in the colonial game.

Under these circumstances, the British and French planners examined numerous alternatives at the grand strategy, strategy, and operational design levels, in choosing the theater of operations and in choosing the main force application thread. The first contingency plan that was considered was to use expeditionary forces in Morocco itself so as to provide direct protection for the vital interests, perhaps with the support of Spain, which also had legitimate interests in Morocco. The second contingency plan was to conduct a naval campaign for control of the western part of the Mediterranean and the Atlantic shipping lanes to Morocco. The third alternative, to place Germany under a naval blockade, was divided into two sub-alternatives: a hermetic, close-to-shore blockade, entailing a high level of friction with the German navy; or a non-hermetic blockade far out at sea, a situation in which Britain's naval superiority would be clearer. The British planners were also of different minds about the blockade's objective: was it a means of drawing out the German navy for a prime major battle, or was it primarily intended to create an economic impact on Germany. A fourth contingency plan was to attack the German navy in port. A fifth was amphibian raids on German targets along the Baltic coast. A sixth was a land engagement on the French–German border in a campaign that might spread to Berlin, Paris, or Brussels. And a seventh alternative involved enlisting Russia for a two-front war. In addition to all of these, there were voices in Britain calling for an eighth option: to abandon the Entente Cordiale with France and instead cut a deal with Germany over Morocco.

The range of alternatives spanned the different objectives of the war, from a localized, restrained solution regarding the ports of Morocco, to undermining Germany's colonial ambitions altogether, and even to the far reaching point of curbing its land power in Europe. Different alternatives shaped a different nature of war, testing different issues: from a test of the military effectiveness on land or at sea, through a test of the ability to enlist allies (Spain, Belgium, Russia), a test of the economic-national stamina (in the case of a naval blockade on Germany), all the way to a test of the ability to enlist national resources (recruiting significant British ground forces to fight a war in Europe, a quick rebuilding of the French navy after its construction according to an erroneous [Jeune École[26]] concept and a German attempt to build a navy that could compete with the British navy). The alternatives ranged from adopting a direct approach to realizing the objectives of the war, to an indirect approach, to an approach so indirect as to make it almost impossible to translate any sort of military end state to the desired strategic–political end state (for example, it is not clear how raids on targets along the German shore would have resulted in a stable end state bringing about the realization of the war's objectives). Every alternative operational theme referred to a different center of gravity, both in essence and in nature, starting with the Tangier port in Morocco, through the shipping lanes, the German economy's dependence on importing resources, the central mass of Germany's ground forces, and key areas in Belgium, to an abstract center of gravity – Germany's strategic fears about a war on two fronts. Some of the operational themes sought to destroy the German army or navy or at least to deal a direct blow against their operating capability, while others sought only to deny Germany its freedom of action on the operational or strategic level. Alternative operational themes brought the sides to an engagement in different places on their efficiency envelopes, reflecting different relative strengths and weaknesses. In some of the alternatives, the sides would have met in the same theater of operations, with or without a prime major battle, while in others each side would have conducted a different parallel campaign, without the two campaigns ever converging in the same theater. There were also ideas on how to deal with the issue of civilian support for the war: the reluctance of the British public and government to commit to a ground campaign in Europe – the continental commitment – but whose support was unquestioning about defending Gibraltar and Belgium.

The wealth of alternatives and implications raises the practical question of how to decide among them, and how to approach the problem of shaping war and planning it. This is where doctrine comes into play, as it is supposed to help us match the military end state to the required political end state, identify the preferred nature of the war, characterize the issues that will be put to the test, define the force application thread, choose the centers of gravity to attack, consolidate the operational theme for attacking the centers of gravity, and only then prepare operational plans. This process of thinking is supposed to reflect our relative strengths and weaknesses in comparison with the enemy's and the circumstances of the particular conflict, and therefore the concept of war cannot be universal. Indeed, though in 1906 the Moroccan crisis dissipated, had a war broken out, each side would presumably have tried to steer it in a different direction, according to its attributes and circumstances. For example, Britain would have tried to direct the war to the high seas, France to a slowly evolving but simultaneous opening of a second front with Russia, and Germany to a quick decision against France in a large ground battle after surprising it through Belgium, and afterwards gaining preparedness for a second front. Doctrine is thus meant to help us answer the subjective national question: what do we do now in the particular war we find ourselves facing?

# 2

# "Simple" Symmetrical Wars

This chapter describes some examples of "simple" wars where the basic assumptions of classical doctrine apply to a large extent. These "simple" wars are characterized, inter alia, by a great deal of symmetry: it is clear that when each of the warring sides seeks to attack its enemy's military-operational center of gravity and sees the prime major battle as the preferred method for attacking the said center of gravity, both warring armies will take up positions in the same theater of operations and for the same purpose – mutual attack of one another's central mass. In these cases, Clausewitz's famous statement generally holds true: that war is difficult to fight but simple to understand.

These wars are also characterized by the fact that they hinge on a single issue, or at least on a central issue – the testing of the sides' military effectiveness – and by the fact that the results at the field levels of the war are directly translated into results on the higher levels of the war. Unlike the wars surveyed in Chapter 3, in these "simple" wars the story of the war is similar at the various levels. These "simple" symmetrical wars are generally characterized, therefore, by their one-dimensional quality: a dominant level (at the field) and a dominant issue (military effectiveness).

The search for prime examples to illustrate the "simple" war is harder than it might seem. In fact, many of the classical wars between regular state adversaries were neither symmetrical nor one-dimensional. A useful place to begin is two of the "big" early Arab–Israeli wars (as opposed to the War of Attrition and the wars against terrorism).

## The Six Day War: The Tactical Field Level as a Dominant Factor

Basic assumptions of classical doctrine to a large extent held true in the 1967 Six Day War on Israel's southern front, the war's main theater. The force buildup of the two sides was similar, and both sides believed that the fate of the war would be sealed at the operational center of gravity located in the operational space, and by attacking it in the prime major battle. The intention of both sides was the meeting of force-on-force in the field, and no attempt was made to act against strategic centers of gravity that did not directly affect the fighting (Operation Moked [Focus] against the Egyptian air force could perhaps be considered a strategic attack, but against a combat element with an immediate and direct effect on the fighting). The campaign was conducted with a clear, direct military objective in mind: to destroy the enemy's assets and inflict a blow to the operational capability of the enemy's fielded formations. The campaign did involve an indirect approach, but mostly at the level of the fielded units.

Despite the vast differences in the sides' circumstances, resources, and stamina, both viewed the war as a test of the military effectiveness of the field levels, with Egypt refraining from exploiting its advantage in terms of stamina. Egypt did not try to undermine the IDF rationale, and did not try to shape a war in which the IDF's advantage would be offset. It "volunteered" to face the IDF with a massive high signature center of gravity, appropriate for attack in a prime major battle, in circumstances, terrain, and deployment ideal for the IDF's relative advantage in joint armor-air battles. Egypt played the role Israel had written for it in its script (as shown in the next chapter, Egypt would not make the same mistake again).

The Egyptians deployed a multi-echelon defense consisting of fortified strong points and mobile reserves in the depth of the Sinai Peninsula. The IDF move toward decision was breaking through the first Egyptian echelon at el-Arish and Rafah, and rapid penetration deep into Sinai to occupy the site at Jebel Livni and Bir Hasna, which denied the Egyptians the opportunity to form a second line of defense. Both sides met in a mobile and symmetrical armored battle, as the Egyptians committed their reserves into counteroffensives, such as, for example, the battle between the 4th Armored Division and the Yoffe Division at Bir Lahfan. The success in this battle contributed to the deep penetration of the Egyptian defense, and to its collapse and inability to fulfill its designated function. Alongside these ground maneuvers, Israel viewed the Egyptian air force as a primary center of gravity, and attacked and destroyed it immediately when the war began (and clearly it too lost its ability to fulfill its designated function).

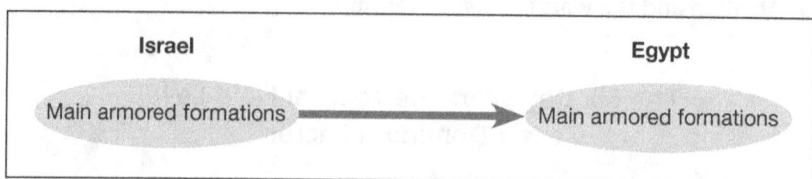

**Figure 2.1** The Six Day War: Both sides take a stand in the same theater of operations for the same objective

Both Israel and Egypt viewed their respective mechanized, armored, and infantry formations as the centers of gravity of their own military forces, and both viewed the enemy's mechanized, armored, and infantry formations as its centers of gravity. Both sides sought to fight a prime major battle in which their armored/mechanized formations would strike at the armor of the enemy.[27] As such, a simple and direct Clausewitzian mirror image of force-on-force emerged (figure 2.1).

The story of the Six Day War is essentially a one-dimensional and "simple" narrative about field levels, especially a story of the tactical level where success on the ground is directly translated to higher levels. There were of course matters at the grand strategy and military strategy levels, such as the decision to call up the reserves and paralyze the economy for an extended period of time, the decision to start a preemptive war, and the decision to attack Egypt first, but these issues do not provide us with a different perspective of the war or with a different outcome.

## The Yom Kippur War, Northern Front: A Mutual Battle of Annihilation between Centers of Mass

In the 1973 Yom Kippur War there was a fascinating split among the Arabs: Syria lagged behind and continued to adhere to the classical, symmetrical Clausewitzian approach, whereas Egypt reached the top league in shaping an asymmetrical, multi-issue, and multi-level war, deconstructing and undermining the rationale of the rival Israeli system. The complexity of the southern front against Egypt is discussed at length in the next chapter, while this chapter points out briefly the classical features of the campaign on the northern front against Syria. Both sides sought to inflict direct and symmetrical blows at one another's centers of mass, which themselves were important centers of gravity.

The Syrian offensive relied on the classical principles of frontal attack, armor-on-armor battles of destruction, and direct, symmetrical, and "simple" force-on-force confrontation. The IDF likewise operated with the rationale of frontal attack against the enemy's center of mass in a prime major battle. After massive destruction of a Syrian force still on Israeli territory, the IDF moved to a counteroffensive on October 10, and on October 11 it broke through to Syrian territory and advanced on the left flank to positions in the range of 40 km from Damascus, from which there was some (symbolic) artillery fire on the outskirts of the city. This created an operational level posture seeming like a threat (that was not carried out) to maneuver towards the Syrian capital. In the face of the massive destruction of the Syrian military mass, the loss of territory, and the pressure aimed at the capital, Syria urgently looked for a way out of the war.

Nonetheless, while the momentum of the IDF offensive on the northern front began to wane, and when the developments on the southern front warranted shifting resources (particularly aerial) and command attention to the main effort on the southern front, the Syrian forces had not yet lost their fighting capability, especially after their reinforcement by expeditionary forces from Iraq and elsewhere. In Clausewitz's terms, the original structure of the Syrian force and its coalition was not yet crushed to the point of being unable to fulfill its designated function, which at this point was limited to securing additional defensive lines deep in the operational space and in the strategic depth of Syria (around Damascus). And indeed, IDF attempts to break through the right flank and its attempts to deepen the success on the left flank were met with stubborn Syrian resistance, which relied on units that neither collapsed nor fell apart.

While some have suggested there was thus no decision against the Syrian forces, one must differentiate between the Syrian forces' defensive capability and their offensive capability. The Syrian forces lost their offensive capability to an extent that enabled Israel to disregard the possibility of another serious Syrian counteroffensive; and based on the rationale underlying the principle of decision in Israeli military doctrine (explained in Chapter 1), it would seem that this is to a large extent the purpose of military decision. The very fact that Israel could decide to change the priorities of air force missions and shift the command attention away from Syria and towards Egypt without taking a significant risk is proof that Syria had in fact lost its offensive capability. This means that in the northern theater of operations, decision was achieved, at least in the terms relevant to the strategic circumstances Israel was facing at the time.

The (partial) decision was thus characterized by the following components. First, the extent of the Syrian forces' destruction denied it offensive capability on the operational and strategic level, and the IDF was able to turn its back on it. Second, Israel demonstrated clear tactical superiority by winning most of the battles and destroying 1,150 Syrian tanks out of an initial deployment of some 1,650, plus destroying approximately 150 more tanks belonging to the Arab expeditionary forces. Israel did, therefore, achieve a significant destruction of mass. Third, Israel took advantage of its combat strength to attain high visibility war achievements, including destroying most of the Syrian armor, breaking through the Syrian forces, and seizing Syrian territory (the enclave and the Syrian part of Mt. Hermon). Fourth and equally important, Israel created an operational level posture seemingly creating a potential threat against Syria's strategic centers of gravity: the general staff reserves, the forces defending Damascus, and of course the Syrian capital itself (although this threat was never carried out, and in light of the exhaustion of the Israeli forces, perhaps the potential never really existed).

The features of the (partial) decision on the northern front during the Yom Kippur War enable a sharper definition of some of the principles of classical doctrine: the decision was achieved despite the enemy's capability of continuing to act with some degree of defensive effectiveness (though not with offensive effectiveness), and it was achieved also by means of a threat against strategic centers of gravity located beyond the operational space in Syria's strategic depth. Israel created a potential for strategic defeat (i.e., a threat against strategic centers of gravity) even before achieving a full military-operational decision (against operational centers of gravity).[28] These circumstances enabled Israel to force an end to the fighting from a position of operational and even strategic dominance, which can be defined as a kind of state-level decision.

The two wars surveyed in this chapter share similar features. First, one may say that both wars are consistent, to a not inconsiderable degree, with the basic assumptions of classical military doctrine as elaborated above.[29] Second, the story of both wars, difficult and casualty-ridden though they were, is a relatively "simple" one: tactical achievements accrued cumulatively to operational level successes, which in turn became strategic. The fielded echelons were at the heart of the story, and the wars hinged primarily on the test of the sides' military effectiveness. Third, the simplicity of the wars – the fact that they were conducted as a symmetrical clash between two military masses in an operational space – enabled Israel in cases such as the campaign against Syria in 1973 to adopt a more restrained decision model. Israel combined proven tactical military superiority with high visibility war accomplishments (such as destroying the forward ranks of the enemy force and territorial conquest), and with creating the potential for a clear classical decision or the potential of a threat against strategic centers of gravity (though this potential was not realized), and thus succeeded in forcing an exit from the war from a dominant military position without translating this dominance into unrestrained military achievements.

Indeed, Israel did not go deeper than the operational space into the enemy's strategic depth in any of its wars (with the exception of attacking obvious military targets located in the depth, such as military airfields and command headquarters); it did not attack the general staff reserves of the enemy or the forces protecting its regime systematically and massively (with the exception of cases in which the enemy itself chose to commit these forces into battle in the operational space); Israel did not attack the enemy's strategic centers of gravity systematically and massively; and Israel never conquered the enemy's capital city (with the exception of Beirut in 1982, an example that must be viewed through the prism of a war on guerilla forces in the territory of a failed host state, and not as part of a "big" war). The main exception is the creation of the potential of an IDF threat against Syria's strategic centers of gravity in 1973, and the potential of this threat was never realized. As noted below, the Arabs too operated in a restrained manner in these wars, and with a few exceptions proving the rule refrained from attacking the Israeli home front. Even during the Yom Kippur War, which was perhaps the most violent of all

## "SIMPLE" SYMMETRICAL WARS

the Arab–Israeli wars to date, both sides behaved with a great deal of restraint (the exchange of strategic fire between Israel and Syria was limited to solitary targets during the 24-hour period of October 8–9, 1973).

# 3

# The Complex Asymmetrical War against a Regular Opponent
## The Picture becomes Multidimensional

This chapter describes and analyzes some of the wars between state opponents that demonstrate the following:

1. Even in a war between state armed forces, different forms and degrees of asymmetry may develop. The more extreme the asymmetry, the less the validity of classical military doctrine. There is no longer a "simple" clash between masses striving to destroy one another in the theater of operations; rather, a multilevel and multidimensional reality is created.
2. When one side tries to shape a test of the sides' military effectiveness, while the other side tries to test a different issue (such as stamina, resources, or the ability to enlist the support of the international community), the validity of the Clausewitzian doctrine declines.
3. In a multilevel war, the struggle at each level hinges on its own autonomous rationale, while the relationships between the various levels in terms of the war's outcome are loosened. The correlation of results from level to level is not direct.
4. Especially true in these complex wars is that realization of the war's objectives depends on a number of wider contexts, and not necessarily only on cumulative tactical or operational level successes against the enemy force's center mass.
5. In such wars, the classical physical definitions of "center of gravity" and "decision" lose their value, and must be endowed with more complex meanings. This chapter attempts to examine some of the analytical tools and terms necessary for dealing with the complex war.

In recent years, the use of the term "asymmetrical war" has become widespread, particularly in the context of confrontations between state forces and non-state or irregular entities. However, almost every war is about looking for one's own relative strength and the enemy's relative weakness; hence, every war involves asymmetry to some degree. Indeed, it is always possible to identify asymmetry whereby the sides have a fundamentally different concept about one of war's major parameters, for example, the buildup of military force and its application, the nature of the specific war in the offing, the military operational rationale that should be adopted, the issue to be tested in the war, the nature of the centers of gravity to be attacked, and so on. Often in such cases the two sides will still meet in the same theater of operations, but not necessarily for the same objective. It would not be an exaggeration to say that most wars are characterized by asymmetry of one kind or another. In this study, asymmetrical war is defined in a broad sense, and the non-state or irregular enemy constitutes just one of its illustrations. The more asymmetrical a war is – especially when each side seeks to reap benefits from the war at different levels or in the context of different issues – the less valid classical doctrine becomes.

## What was Tested in World War II?

An example of asymmetry in the nature of the war in terms of complexity and multidimensionality may be found in the German campaign against France (1940). The French prepared for a follow-up campaign to World War I, and therefore expected it to be defensive, protracted, and static in nature, and to rely on the fortifications, obstacles, and firepower of the Maginot Line. This would be a campaign testing the stamina and resources of the two sides. By contrast, the Germans prepared for a campaign that would overcome the difficulties of achieving decision that emerged in World War I, and therefore its nature would be offensive, short, and dynamic, relying on the rapid flanking maneuver (blitzkrieg); the fate of the campaign would be sealed by military effectiveness primarily at the operational level. It is possible to go even further and claim that the sides positioned themselves in the theater of operations with different objectives in mind, because the Germans simply did not take up positions for the exhausting trench war for which the French had prepared.

The Germans did not seek to attack the Maginot Line directly, rather to attack the French war plans and paradigms. They made a rapid attack on the far right flank (beyond the Maginot Line) deep into Belgian and Dutch territory, which prompted the commitment of French reserves and British expeditionary forces through northern France deep into Belgium. The Germans then attacked the right flank a second time, though a little closer to the Maginot Line, through the Ardennes Forest, which enabled them to encircle the French–British forces in northern France and Belgium and sever them from the Maginot Line. They were also able to bypass the Maginot Line itself with relative ease because the mobile reserves positioned behind it were already committed and trapped in Dunkirk. The German maneuver created a new French–British center of mass in a location favored by the Germans, northern France and Belgium, while in practice deconstructing the rationale of the Maginot Line: the first (far) move prompted the commitment of the mobile reserves towards Belgium, leaving the Maginot Line as an abandoned obstacle that once bypassed was rendered ineffective. The second (near) move severed the reserves from the heart of France and denied them the capability of acting to defend it. This maneuver caused the French defenses to lose their operational level posture and their capability of acting effectively to realize their designated function, bringing about the decision in the campaign. The decision did not stem from the destruction of mass or from the symmetrical clash of two masses, rather mainly from deconstructing the enemy's rationale.

The 1940 campaign in France also demonstrates the differences in the story of the war at its various levels. The Germans indeed achieved an unequivocal decision at the operational level, but at the techno-tactical, tactical, and logistical levels they were on the brink of failure. At that time, only a few Wehrmacht units were properly equipped for blitzkrieg, and when the few armored outfits rapidly surged ahead, the infantry, artillery, and logistics trailed dozens of kilometers behind. The armored units moved forward without leaving forces behind to secure the flank and the rear, and without establishing logistical supply lines. At times, the tank engines were silenced for hours or even days due to a lack of fuel. Because of the shortage of main battle tanks, in certain locations the German lines were made up of motorcycles, trucks, and other soft vehicles lacking serious fighting capability. Nonetheless, these difficulties and logistical, techno-tactical and tactical failures did not overshadow the Germans' capability of creating an operational level posture undermining the French and achieving an operational level success. The disparity between the operational level success and the less successful outcome at the lower levels is apparent.

This holds true for the higher levels as well: the characteristics of World War II at the military strategy and grand strategy levels were such that the Germans did not succeed in translating operational level successes – such as the one against France in 1940 – into victory. A full analysis of the complexity of World War II is beyond the scope of this study,[30] but a few concise points can be added. At the grand strategic level, Germany chose to go to war against what at the time were the three strongest countries in the world – the United States, the Soviet Union, and Britain. Germany "succeeded" in uniting the two Western powers with the Soviet Union despite the vast gulf separating them (until the outbreak of the war, many in the West viewed the Soviet Union as more dangerous than Nazi

Germany). Churchill himself thought that had Germany refrained from attacking Britain, attacking American ships in the Atlantic, and violating the Ribbentrop-Molotov Pact, it would have been capable of successfully defending its very significant achievements in western and central Europe. Indeed, at the grand strategy level, Germany did not set itself limited and seemingly legitimate objectives (such as an end to the Treaty of Versailles or the recovery of German territory lost over previous decades) that the superpowers could have tolerated. However, Germany did not have the national resources, stamina, and staying power required for a protracted simultaneous conflict with three superpowers. Moreover, Germany chose allies whose contribution was limited: Italy did not represent a significant force (the war production of the Ford company alone was greater than that of all of Italy), and Japan, though engaged in fighting against the United States and Britain, did not do so in a synergistic effort. Furthermore, the combined resources of the Axis nations never represented a substantive challenge to the Allies. In 1941, for example, the United States produced more steel, aluminum, oil, and vehicles than the rest of the world put together, and in 1941–45, American production doubled.

Beyond the reality of the sides' basic resources, the Allies had an advantage in enlisting resources and placing them at the disposal of the war effort. The United States converted its peacetime industry into wartime industry very rapidly, while the Soviet Union succeeded in moving 16 million workers and 2,500 essential factories east of their conquered territory. Moreover, the Allies disrupted the flow of natural resources to the members of the Axis, and their air forces damaged the military industry of both Germany and Japan.[31] The German grand strategy also suffered from a lack of synchronization between the military effort and the capabilities of the military industry. In 1939, the German war industry was just starting and the industrial capabilities reached their peak in 1943–44, only after a great deal of erosion in German manpower and the ability to transport weapons to the front – in fact, too late to have an impact on the course of the war. Germany sought to test the military effectiveness of the sides, but found itself deeply involved in a few capital-intensive campaigns against the US and Britain, and in a manpower-intensive campaign against the Soviet Union.

A similar picture emerged in the Pacific: in the first big battle of the campaign, at Midway, four Japanese aircraft carriers fought three American carriers. In the two years following the battle of Midway, Japan built seven more aircraft carriers, while the US built another 90. In 1940, Japan built 30 large naval vessels, and the US built none at all. However, in 1943, Japan expanded its output to 122 vessels while the United States produced 2,654 large vessels (some intended for the European theater). In light of these overwhelming disparities, even if the United States had been defeated at Midway, the Coral Islands, and Guadalcanal, and even had it been defeated in every naval battle in 1942 and 1943, in the end, after another month or another year, the asymmetry in resources would have sealed the fate of the campaign in the Pacific Ocean. Japan could not have withstood the steamroller of American industry. And indeed, the asymmetry in war industry output generated facts with almost inevitable significance.[32]

On the military strategy level, Germany embarked on too many campaigns of too far reaching proportions. The classical Prussian way of war was to use maximum force to attain quick military decision, but in the service of restrained political objectives and as preparation for a negotiated and balanced peace. However, this time the German military found itself overstretching in many military efforts without any practical end point, reaching from the Caucasus Mountains to the shores of Normandy, and from the Balkans and the Saharan Desert to the fjords of Norway. In some of the campaigns, such as against the Soviet Union and even in North Africa, it was not possible to point to stable, lasting end states. Even in the face of repeated successes at the field level and the capturing of additional territories, these campaigns were destined to continue until the Wehrmacht's exhaustion and logistical breakdown and the collapse of the German war machine. German military effectiveness dissipated primarily because of the vast expanses and the Soviet stubborn resistance, defiance, and capability of recovery.

Furthermore, in some aspects, even on the military strategy and operational levels, World War II seems like a war of resources. For example, both the Axis and the Allies recognized the centrality of controlling the sea, as raw materials, weapons, and fighters were all transported using the shipping lanes, and this was the only way to transport the American military mass to the occupied European continent. Control of the sea was therefore a major force application thread. However, the campaign for the Atlantic Ocean itself was determined by the pace of the application and destruction of resources, and not by managing large scale, brilliant battles. It was actually a confrontation involving the development of four statistical factors: the rate of sinking cargo ships, the rate of sinking submarines, the rate of building cargo ships, and the rate of building submarines.[33]

Thus in the face of such fundamental inferiority at the military strategy and grand strategy levels, Germany could not translate a series of operational level successes into victory. One could even state the opposite: the more Germany accrued operational level successes, the more it stretched itself over larger areas, assumed a growing number of missions, and thinned out its forces on every front. The logistical burden grew, and this in turn exacerbated Germany's inferiority at the higher levels. As on the lower levels, the higher levels of war likewise display autonomous distinct rationale at every level, and correlations between their outcomes are more tenuous. In the end, the war hinged on issues of national resources and staying power, and therefore those centers of gravity whose collapse sealed the war's fate were not operational, like the Maginot Line or Midway Island, rather strategic, like sources of raw materials, naval supply lines, and the military industries (in fact, these centers of gravity belonged to the grand strategy level, but their common appellation, in use also in this study, is strategic centers of gravity). Other critical strategic centers of gravity were related to the attrition of the Wehrmacht, primarily as a direct result of its attempt to conquer the vast reaches of the Soviet Union. These centers of gravity belong to the military strategy level and are abstract, even if the way to them passed through physical-operational milestones such as Kursk and Stalingrad. The Allies achieved a total military decision in the war, and towards the end of the war, their military effectiveness improved beyond recognition. However, this decision did not emerge merely from a contest of military effectiveness, but was achieved at the end by a prolonged resources and stamina race. By contrast, what characterized the years 1939–43 was the inability of the Axis nations to translate their impressive operational level successes into a military decision in the war as a whole and a long enough time passed without a military decision, so that other issues – the enlisting and placement of resources and stamina – became dominant and sealed the fate of the war.

## The Yom Kippur War, Southern Front: Egypt Deconstructs the Israeli Paradigm

In Israel's "big" wars against regular state armed forces, it is also possible to discern signs of emerging asymmetry, complexity, and multi-dimensionality, first seen on the southern front in the Yom Kippur War (1973). The last time that Egypt deployed for a classical symmetrical war testing the field effectiveness of the sides was in the Six Day War, and the last Arab party to have initiated a classical symmetrical war with Israel was Syria in 1973. Since then, the Arabs have tried to shape and force on Israel wars testing the staying power and stamina of the sides, and the ability to enlist the international community to their cause. The model for challenging stamina changed from time to time and became more sophisticated, from the War of Attrition through the intifada, terrorism, and guerilla warfare, to wars of rockets and missiles, to adjustment to fighting against RMA capabilities.

However, the transition from wars that symmetrically test the field effectiveness of the sides to more complex wars is discernible in "big" wars as well, in which the absolute superiority demonstrated by Israel's strike forces (the armored corps and the air force) in the Six Day War, along with recognition of the general ineffectiveness of the Egyptian armed forces, prompted Egypt to seek asymmetrical and indirect ways of attaining its objectives.[34]

Immediately after the Six Day War, Nasser coined the phrase, "That which was taken by force shall be restored only by force." The plan was for Egypt to retake the Sinai Peninsula (and to conquer territory even beyond it), while at the same time destroying IDF forces deployed in the theater of operations in a prime major battle; thus the main focus of Egyptian activity lay in building up the military force so that it would be able to fulfill this designated function in the future. However, from experience accrued during the War of Attrition and the emergence of a new generation of key personnel – such as President Sadat, Chief of Staff Shazly, and Chief of Operations Gamasi – Egyptian thinking gradually underwent a change. Instead of the starting point being an ambitious war plan, with attention paid to force buildup to realize the plan, Egypt now sought to go to war at once. The starting point was the existing capabilities and forces, and Egypt looked for paradigms that would enable it to achieve its aims – even with the limited field capabilities at its disposal. The Egyptian military understood that it was not capable of undertaking operational level maneuvers, combined arms cooperation, and dynamic management of an evolving campaign. Therefore, it sought to undertake a limited, static and well planned move. The Egyptians also sought to offset the IDF's armor and air superiority through anti-tank and surface-to-air missiles, but these forces (especially the SAMs) were not suited in those years to rapid and deep mobile warfare. Egypt could therefore undertake a limited move only, by advancing to the outskirts of the area protected by SAMs (Integrated Air Defense System) – made up of batteries that remained on the Egyptian bank of the canal, deploy anti-tank forces at the edges of the Integrated Air Defense System, and move into a defensive posture while waiting for the IDF counterattack. The question was how to use these limited field capabilities to enable Egypt to act effectively to realize its political objectives. The answer required a thorough deconstruction of Israel's plans, assumptions, paradigms, and approaches as well as the development of a response for them at each level of the war. To explore this requires a survey of the war, level by level.

On the level of political and national objectives, Egypt sought to challenge the status quo and restore the Sinai and the Suez Canal to its control. It aimed to cross over to the American bloc and receive economic and military aid from the United States, and it sought to restore the prestige of the regime and the country and erase the humiliation of 1967. Israel, surprised by the outbreak of the war, only tried to deny Egypt of war achievements.

At the grand strategy level, Israel assumed that the series of military decisions it had accrued would create grand deterrence, i.e., deterrence not dependent on the context of a particular conflict, but that would channel

Arab grand strategy on the whole in some other direction, because no Arab nation would dare pay the probable price of war. Israel was also relying on its close, intimate alliance with the United States. For these two reasons, it assumed that the status quo resulting from the Six Day War had produced a stable situation, and that sooner or later Egypt would be forced to accept a political compromise on Israel's terms (including the partitioning of the Sinai). The beginning of the Egyptian response lay in rejecting the idea of deterrence, with Sadat willing to pay a tremendous cost in the war's loss of life and resources.[35] Sadat himself said that the Egyptian approach was meant to convince Israel that a defense doctrine based on deterrence "is not a steel shield that cannot be breached."[36] Egypt also estimated that the most effective way of propelling Israel to retreat from the Sinai would be through American pressure, and therefore sought to push the United States in that direction. The idea was to demonstrate to the United States that Israel's conquest of the Sinai had created an unstable reality, and that an Egypt-Israel war might threaten the relationships between the superpowers and endanger the détente. Egypt also sought to demonstrate to the United States that an Egyptian–Israeli flare-up would reverberate in the American and global economy via the oil weapon. Sadat assumed that the American and the Israeli perspectives over the costs and risks that an Israeli victory was worth were incongruent (this notion ultimately proved true when, for example, the US opposed Israel's encirclement of the Egyptian 3rd Army and then opposed its destruction, and insisted on bringing in supplies to the besieged Egyptians). Egypt also wanted to demonstrate to the public-political system in Israel that the status quo was unstable and that its perpetuation would entail an intolerable price to the Israeli public in casualties and would also shut down the economy as a result of mobilizing the reserves. Egypt sought to undermine the cohesion of Israel's government–military–civilian triangle ("the wondrous trinity," to cite Clausewitz), and strike a blow at Israel's self-confidence.

At the military strategy level, Israel saw the Sinai as providing strategic depth that if crossed would leave the Egyptians exposed as they moved through the desert expanses. Israel relied on a concept of an early warning, a defensive blocking stage using a small regular force, then mobilizing its reservists, and launching counteroffensives. It strove to achieve a sequential decision among the fronts, in which as in all previous wars it would defeat the strongest enemy – Egypt – first (the "Egypt first" concept). In addition, Israel had generally sought to craft short wars testing the sides' military effectiveness. The Egyptian response to Israel's strategy was complex, partly at the military strategy level and partly at the lower levels. First, a change in the Egyptian war paradigm rendered Sinai's strategic depth less relevant (Egypt had no intention of crossing the desert), also making it difficult to receive any early warning, both on the strategic level and on the operational and tactical levels.[37] In response to the Israeli desire for a sequential decision with "Egypt first," Egypt sought to attack Israel simultaneously with Syria, and encouraged Syria to adopt a much more aggressive, intensive, and threatening approach than Egypt's, so as to direct Israel towards a "Syria first" strategy. The more important strategic deception perpetrated by Egypt was not directed against Israel, but against its ally, Syria, misleading it with regard to Egypt's planned mode of fighting. Egypt was interested in shaping a war testing the stamina and staying power of the sides as well as their ability to enlist the support of the international community, while at the same time denying Israel an opportunity for decision and the option of a unilateral military exit from the war. It sought a quick completion to the offensive stage of the war against the IDF's small regular force, and a move to a static defensive posture before the arrival of Israel's reserves. The Egyptian strategy sought only limited territorial military success (without penetrating the depth of the Sinai), which would be defended until international intervention and an imposed ceasefire.

The analysis of the war at the operational level must be divided into two stages: until October 14, with Egypt adhering to a successful mode of action, at least according to its own definitions, and after October 14, when Egypt erred and enabled the IDF to turn the tables at the operational level. The IDF's operational-systemic working assumption was that Egypt would engage in an armored-mechanized attack in several main efforts to the depth of Sinai. That kind of "heavy" action would afford Israel both the time required to organize and the opportunity for a decision in combined arms mobile warfare: a heavy Egyptian attack through the open expanses of the Sinai would create ideal conditions for the armored corps and the air force to destroy the Egyptian center of mass in a prime major battle. The IDF also assumed that this Egyptian mode of operation would enable Israel's air force to operate effectively against Egypt's heavy bridging equipment and bridgeheads.

The Egyptian operational level response was merely to engage in a limited bite-and-hold operation along the width of the front without any one main effort. This idea was designed to present the IDF with a dilemma: to concentrate its forces for a major effort and attain success only on a limited part of the front, or to divide up its forces and use them ineffectively. And indeed, in the first stage, the Egyptian infantry crossed along the entire width of the canal using about 1,000 light crossing means that did not present suitable targets for the Israeli air force. The infantry crossing stage was completed almost entirely within three hours, and the Egyptian infantry adopted a defensive posture before Israel managed to create a significant counterattack (and certainly before the arrival of the reserves). It was only afterwards that the Egyptian armor and other heavy equipment started crossing the canal. The forces that crossed quickly adopted a static defensive posture while simultaneously deploying a tight formation of anti-tank means; all Egyptian forces that crossed to the eastern bank stayed close to the canal (within approximately 8 km), and enjoyed cover from surface-to-air and artillery batteries that remained in place on the western bank of the canal. The moment this static deployment was in place, the only achievement required of the Egyptian force was to survive as a functioning defensive formation on the eastern bank, causing a maximum of casualties to IDF counterattacks and demonstrating the excessive cost of routing out the Egyptian forces. The Egyptian military did not really engage in an offensive, rather just advanced its defensive line from the western bank of the Suez Canal to the eastern bank. And, indeed, this first stage of the campaign succeeded beyond expectations, as the IDF made several systemic-operational errors, such as a rigid defense of its outposts ("maozim") and scattered tactical counteroffensives with extremely small forces.

Starting on October 14, Egypt made two critical blunders at the operational level. First, it moved some of its general staff reserves and field armies reserves from the canal's western bank to the eastern bank. In doing so, it thinned out the defensive line on the western bank, and the overall Egyptian defense formation lost some of its depth and a significant portion of its mobile reserves. Second, the Egyptian military deviated from the defensive-static mode and embarked on a large mobile armored offensive on open ground outside the range of its surface-to-air missiles and artillery batteries, most of which remained on the western bank of the canal. The IDF exploited these two errors effectively: it used the Egyptian offense to conduct the biggest armor-on-armor battle since World War II, and achieved significant destruction of the mass of Egyptian forces. The IDF identified a seam between the two Egyptian field armies, used it to cross the canal, and took advantage of the thinned-out Egyptian forces on the western bank and the weakening of the mobile reserves to rapidly expand the bridgehead and break through from it. The Egyptian 3rd Army was surrounded, the Egyptian 2nd Army's supply lines were threatened, and the IDF assumed positions on the major roads leading from the canal towards Cairo, with only a reduced Egyptian defense formation standing between it and the Egyptian capital. It is doubtful whether the weary IDF was capable at that stage of breaking through to the west towards Cairo or destroying the 2nd Army, but its operational level posture certainly created that impression. On the opposing side, the Egyptian military lost its posture and capability at the operational level: first of all, the extent of destruction to the Egyptian mass caused the loss of offensive capability; second, there were significant cracks and breaches in the Egyptian lines so that they lost both continuity and depth, the mobile defense in the depth was thinned out, significant forces were either destroyed or surrounded, and as a front, it had difficulty fulfilling even its defensive function.

At the tactical level, the IDF relied on an array of outposts along the length of the canal and a second line of strongholds ("taozim") approximately 20 km behind the outposts. On the eve of the war, only one single division scattered its forces along the width of the entire front and into the depth between the two arrays. The IDF's war plans relied on a variant of the blitzkrieg – a rapid maneuver of armored forces operating jointly with the air force serving as flying artillery, with a preference for flanking rather than frontal assaults. The Egyptian tactical response began with a change in perception about the maozim outposts. In the years leading up to the war, Israel had closed down many of the outposts and reduced the forces stationed in those remaining. Therefore, the Egyptians correctly noted that the spaces between the manned Israeli outposts were so large as to render them ineffective as a defensive system providing continuity of fire and mutual cover.[38] Egypt decided not to deal with the outposts directly at the first stage of the war, rather to cross at the gaps between them, isolate them, exhaust them, and attack them from the rear. The Egyptians exploited the small size of the Israeli front line force at the water's edge and the element of surprise to create a huge advantage in numbers. Each Egyptian force crossed the canal from its defensive position, without significant sideward movements, limiting its exposure to fire. The Egyptian movement extended to the shortest distance possible, preferring dense deployment and saturating the line with forces and fire over creating depth along the eastern bank. The Egyptians acted to link the bridgeheads rapidly in order to create a continuous deployment along the width of the entire front, with no flank. The Egyptian tactical plans were successful until the operational level reversal of October 14, when Israel regained tactical superiority as well. Once again, the IDF engaged in massing its forces for a main effort, exploited its success in advancing rapidly, and relied on the joint fighting of the armored corps, the infantry, the artillery, and the air force. After breaking through the Egyptian forces, the IDF returned to its preferred tactic of attacking the enemy from its soft rear, instead of engaging in a frontal assault.

At the techno-tactical level, the IDF relied on aircraft and tanks, while its front line also relied on the obstacle of the Suez Canal and the high rampart it had constructed along it. The Egyptian forces responded to the challenge of tanks with a combined anti-tank blend, including anti-tank cannons, anti-tank missiles, grenade launchers, and anti-tank mines. Anti-tank measures shifted from being a capability reserved for designated units to an organic capability of every unit. As for aircraft, the Egyptians deployed surface-to-air missiles (SAMs) and anti-aircraft artillery (AAA), and in place of close air support they used a saturated artillery formation. The shift in the Egyptian war paradigm created an additional techno-tactical quandary: a significant portion of Israeli armaments (both aerial and tank shells) was meant to be used against tanks and hard targets, and was unsuitable for use against infantry forces. As for the obstacle, the Egyptian response was constructed of a variety of crossing methods and engineering solutions to break through the rampart, such as high pressure water cannons. Towards the end of the war, the IDF adjusted to the Egyptian techno-tactical challenge, and inter alia, began to attack the SAM batteries deployed on the western bank of the Suez canal using tanks, while the air force struck at the Egyptian armor operating outside the umbrella of the Integrated Air Defense System.

Table 3.1 summarizes the different war levels on the Yom Kippur War's southern front.

**Table 3.1** Egyptian–Israeli Battle of Rationales (Yom Kippur War, 1973)

| Level of the war | Israeli plans/assumptions | Egyptian attack on Israeli plans/assumptions | Egyptian success |
|---|---|---|---|
| **Grand strategy** | A series of military decisions accumulating to grand deterrence; a close alliance with the United States; a stable status quo; sooner or later, Egypt would despair of the military option and accept a political compromise on Israeli terms. | Egypt would not be deterred, even if paid a high price; only the US could make Israel withdraw; demonstrating to the US that the status quo was not stable and threatened the global détente and economy; an Israeli victory would cost the US dearly; demonstrating to Israeli civilians that the status quo entailed an intolerable price; undermining the self-confidence of Israel and the cohesion between the government, the military, and the civilians; limited military success would translate into complete political success. | Fully realized |
| **Military strategy** | Sinai as strategic depth; a small regular force (and the air force) for a temporary blocking stage and serving as an early warning system allowing mobilization; a clear military decision, front after front, with "Egypt first"; short wars testing the sides' strike capabilities. | A shift in paradigm rendering the Sinai depth less relevant and making early warning more difficult; a simultaneous Egyptian-Syrian surprise attack (with limited Soviet backing); pushing Syria to adopt a more aggressive attitude than Egypt (leading to a "Syria first" policy); testing the stamina of the sides and denying Israel the opportunity for decision; international intervention and a ceasefire protecting a limited territorial gain. | Realized to a large extent |

## THE PICTURE BECOMES MULTIDIMENSIONAL

| | | | |
|---|---|---|---|
| Operational (until October 14) | An assumption that the Egyptian assualt would progress with a few main efforts into the depth of the Sinai; a plan to concentrate efforts to counter Egyptian maneuvering efforts into the Sinai depth, but in practice using small forces and scattered efforts; a rigid defense along the canal based on the outposts; a plan to destroy bridges using the air force. | Opening with a limited infantry bite-and-hold along the entire canal (using 1,000 light crossing means, without any one single major effort), completed before the arrival of IDF reserve forces; limited advance inside of the Integrated Air Defense System coverage and halting, deploying dense anti-tank defenses; absorbing counterattacks; causing casualties among the IDF; surviving as a functioning defense line. | Realized to a large extent |
| Operational (from October 14)* | A main effort in an armor-on-armor battle and massive destruction of the enemy; crossing to the western bank of the canal in the seam between the Egyptian armies; surrounding the 3rd Army and threatening the supply lines of the 2nd Army; demonstrating presence on the road to Cairo. | Moving the reserves from west of the canal to the east, weakening the depth of the defense system, and weakening the defenses of the western bank and the road to Cairo; a large armor-on-armor battle, with movement on the open ground outside the umbrella of the Integrated Air Defense System and the anti-tank formations. | Egyptian failure, and a military decision for Israel. |
| Tactical | At the outset, defense in the outposts; scattered frontal counterattacks by small armored forces.<br><br>Later, a main effort in mobile warfare and flanking maneuvers, using combined forces of armor-infantry-artillery-air. | Understanding that the outposts were not a defensive line (bypassing, isolating, exhausting, and attacking from the rear); taking advantage of the element of surprise to gain an advantage in numbers against the frontline; each force crossing from its defensive position, without widthwise movement or prior changes in deployment; entrenchment of anti-tank defenses and mines; decreasing the depth of the bridgeheads enables greater density of anti-tank assets and firepower; a continuous defense without a flank. | Realized to a partial extent, at the end – tactical inferiority. |

| Techno-tactical | Focus on tanks and aircraft; much weight given to armaments not suited for use against infantry; the canal, the earthen rampart, and other obstacles. | Multi-layer blended anti-tank measures. Shifting anti-tank measures away from only designated units to an organic capability; SAMs against aircraft; concentrated airtillery as a substitute for air support; water cannons and other engineering equipment for contending with the obstacles. | Realized to a large extent at the beginning, less so as the IDF adapted progressively better |
|---|---|---|---|
| | Later, armored units against SAMs, and the air force against armor. | | |

(*) At this level, the Egyptian moves should be read first, and then Israel's taking advantage of the opportunities.

What then was the outcome of the war? The answer depends on the particular level examined. Egypt attained the political goals it had set for itself, and thus may be said to have won the war. Egypt also to a large extent realized its plans at the grand strategy and military strategy levels. However, at the operational level, the IDF gained a military decision: the Egyptian military lost its ability to fulfill its designated offensive function and even lost its defensive functionality as a front, and the IDF's operational level posture pierced the Egyptian defenses in two, penetrated them almost to their full depth, and seemed to demonstrate the potential for executing a maneuver towards the capital. Nonetheless, despite the IDF's tactical superiority, at the military tactical level the Egyptian armed forces did not collapse. The 2nd Army continued to fulfill its designated defensive function, and most of the Egyptian divisions continued to defend their respective areas of responsibility with some level of effectiveness. At the techno-tactical level, Israel ultimately emerged having the upper hand, though by a smaller margin.

The tension in the outcome of the war on its different levels emerges from an argument between President Sadat and Chief of Staff Shazly about how to respond to the IDF's westward crossing of the Suez Canal.[39] Shazly was focused on the operational level and saw before him a front that had lost its continuity, had had its depth penetrated, and whose mobile reserves in the rear were thinned out. He therefore requested that significant forces be returned from the eastern bank of the canal to the western bank and that Egyptian operational level posture be restored. Sadat, however, eyed the situation as a whole: he understood that the rising tension between the United States and the Soviet Union would impel the Americans to restrain Israel and even bring about a rapid, imposed cease-fire. He therefore was less concerned about the Israelis deepening their operational level success or about the potential implications at the higher levels. He estimated – apparently correctly – that only a significant

Egyptian presence on the eastern bank during a ceasefire could move the political process forward that would result in restoring the Sinai to Egypt, and therefore he refused to withdraw any force whatsoever from the eastern bank ("not one solider or one rifle").

Using military means, Egypt, a non-industrialized country with one of the highest illiteracy rates in the world, attained its political wishes from the industrialized, technologically advanced State of Israel (i.e., it won). It did not do so through a classical, symmetrical confrontation of force-on-force between two fielded formations, as the "generic" Egyptian military effectiveness was low. Egypt's success lay in taking a broad view of the war – a much broader view than the prism of weapon systems or the tactical units – and in understanding the situation as a whole, as well as in choosing correct grand strategy and military strategy, which depended little on the outcomes at the operational, tactical, and techno-tactical levels. While classical wars such as the Napoleonic wars or World War II necessitated waiting until the military end state was clarified before it was possible to shape the political end state, Egypt sought first and foremost to deal with shaping the political–strategic end state while the military moves at the field levels were only meant to upset the equilibrium and provide a catalyst for moving towards the political end state. Indeed, the political–strategic end state did not emerge as a direct result of the military-operational level end state. Moreover, even if it was not explicitly understood as such, in practice Egypt was sowing the seeds for arranging a political end state years before the war, in a dialogue it conducted with the United States about the William Rogers and Henry Kissinger plans, as well as with UN Special Representative Gunnar Jarring. It may not be possible to claim that Egypt was on solid political ground, but it certainly identified and shaped the realistic outline of a political end state long before the outbreak of the war. The fact that Israel tried to defend a status quo that was seen as illegitimate, i.e., not agreed to by the international political community, and caused the closing of the Suez Canal to shipping, whereas Egypt was attempting to make the transition to a different reality, understanding that this was something the international community could accept, enabled Egypt to arrive at a political end state by way of only a military shock and a demonstration of the status quo's instability.

Despite the self-evident interrelationships between the different levels of the war, the outcome of the war at each level differs in an essential way (figure 3.1). That is, the levels not only feed one another (up and down), but they are to a large extent also autonomous and isolated from one another, and a success or a failure at one level does not necessarily generate a corresponding outcome at another. What is particularly striking is that despite the Egyptian error at the operational level on October 14 – the error that enabled Israel to turn the tables and achieve a military decision – the outcome of the war at the higher levels remained almost unchanged. The

# THE COMPLEX ASYMMETRICAL WAR AGAINST A REGULAR OPPONENT

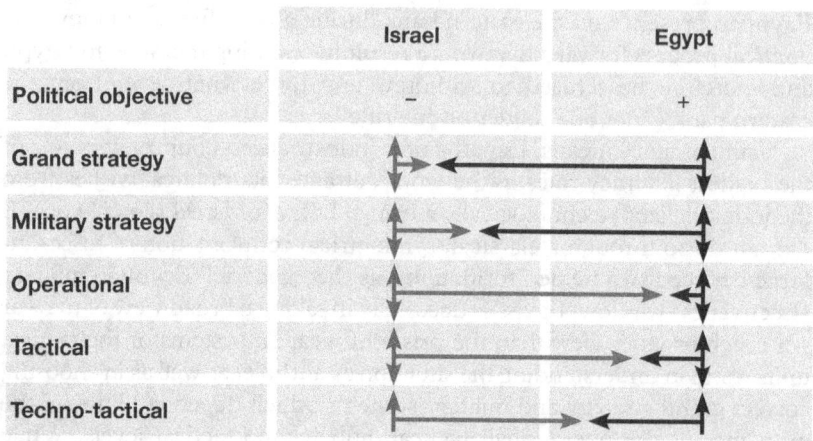

**Figure 3.1** The Yom Kippur War, Southern Front: Different levels, different outcomes

tactical and operational level blows to the Egyptian military's capability did not deny Egypt the ability to erode Israel's stamina, to undermine the cohesion between government, military, and civilians, to persuade the civilians in Israel that the status quo was unstable, and to demonstrate that Israeli deterrence is not a steel shield. Despite the siege of the Egyptian 3rd Army, the blow to the 2nd Army, and the IDF's positioning on the major roads leading to Cairo, Egypt succeeded in demonstrating to the United States that Israel's occupation of the Sinai was a threat to détente and to the US and global economies, and to create a divide between American and Israeli interests. Despite the IDF's tactical and operational superiority, Egypt still succeeded in initiating a political process that would restore the Sinai to its control, and in exchanging its Soviet patron for the US, thereby gaining critically needed American economic aid and the opportunity to rebuild its armed forces with Western means of warfare. Thus correlation between the outcome at the lower levels and even at the operational level to the outcome at the higher levels was weak if at all existent. The Israeli military decision at the operational level was not translated into successes at higher levels. The Egyptian genius was manifested precisely in that: the ability to shape a war whose circumstances and paradigm would attain the objectives effectively, despite the "generic" ineffectiveness of the Egyptian forces.

Egypt succeeded in inventing an interesting hybrid: the offensive defense. It attacked Israel at the grand strategy and strategy levels, but within a few hours, the Egyptian armed forces adopted a defensive mode at the operational[40] and tactical levels, and maintained that mode until the end of the fighting (table 3.2). This offensive defense fully exploited

Egypt's advantage and Israel's weakness. It enabled Egypt to conduct the fielded battle on its own terms, forcing the IDF to attack the dense defensive formation frontally and sustain many casualties, and with the apparent ability to extend the conflict at will, as loss of life and a paralysis of the economy over time are Israel's primary weaknesses.

**Table 3.2** Change in the Egyptian Paradigm at the Operational Level

|     |     | Egyptian armed forces | |
|     |     | Maneuver | Static defense/fire only |
| --- | --- | --- | --- |
| IDF | Maneuver | Symmetrical decision campaign (assessment of IDF Military Intelligence and the Egyptian plan of 1968) ⇨ | Strategic offense and operational defense (moving the defensive line to the eastern side of the canal; the Egyptian plan of 1973) |
|     | Static defense/ fire only |  | The War of Attrition (1969–70) |

The Egyptian war paradigm provides additional perspectives to classical Clausewitz doctrine in several ways and produces a set of complementary analytical tools and concepts. First, it is possible to test how the Yom Kippur War's southern front provides perspective on the Clausewitzian concept of decision. On the one hand, Egypt sought to win the Yom Kippur War (namely, to realize the political objective) even without a military decision. It is clear that Egypt had no intention of conducting mobile warfare deep into the Israeli lines and had no intention of trying to destroy the mass of the IDF's fighting force or cause the collapse of its original structure. On the other hand, Egypt nonetheless sought to impede the IDF's capability of operating effectively against the Egyptian military and to fulfill its designated function, even if this was based more on a shift in the nature and rationale of action by the Egyptian military itself and less on the direct destruction of the Israeli fighting force. In this sense, Egypt indeed presumed to try to force a paradigm of war in which from the outset the IDF's capabilities would not have the chance to be manifested effectively, and this situation echoes the classical definition of military decision. In other words, decision does not deal only with the question of which side succeeds in destroying the mass of enemy forces in the course of the fighting, but also with which side builds up its force, makes war plans, and imposes a nature and rationale of war that will render that side effective in war and the other side ineffective from the very start. When one side manages to force a paradigm of war in which that side is effective and the opposing side is ineffective, the war has almost been decided even before the first shot is fired. In such a case, the period of time that affects military decision most is the period of building up the military force and shaping the paradigm of the war, namely before the war, and not the period when the battle is underway.

Indeed, in 1973 Egypt sought to shape a systemic operational design in which the IDF's relative advantage would be less relevant. So, for example, the IDF buildup of its forces and its war plans to concentrate aerial and armored efforts against heavy armored efforts that penetrate into the depths of the Sinai were less relevant in the case of infantry crossing the entire width of the Suez Canal, advancing only within the protective umbrella of the Integrated Air Defense System and anti-tank defenses. Similarly, the Maginot Line was irrelevant against the German blitz penetration from Belgium, the Wehrmacht was less relevant in an extended conquest in the depth of Russian territory, American armored divisions were irrelevant against the Vietcong, and light cavalry was irrelevant in a trench war.

When there is asymmetry in the buildup of forces and in their operational concept, and each side prepares its forces for a different war, the real war hinges on the question of which side will be able to force the war it has planned. If one side is trained to play basketball and the other to play chess, the critical question is whether the game will be decided by a slam dunk or by a checkmate. A force that trains to slam dunk but finds a board with 64 squares and 32 black and white pieces in the theater of operations is from the outset not equipped with the relevant capability to fulfill its designated function, and is almost ineffective, as if the decision has occurred. Therefore, one of the core issues of war is predicting – and, even more importantly, successfully shaping and forcing – the rules of the game, namely the nature of the war, its rationale, and the issues on which it hinges. At first it was Egypt that succeeded in dictating the nature of the war it had prepared for, and therefore the relevance and the capabilities of the IDF to realize its plans and function were impaired. After the reversal of October 14, the nature of the war at the operational and tactical levels changed, and the IDF once again became relevant, whereas the Egyptian military lost its capability to act with relevance and effectiveness in the new context. Nonetheless, even the IDF's renewed operational effectiveness after October 14 was not relevant to the higher levels of the war. For example, its field capabilities were still not able to obstruct Egypt's plan to demonstrate to the Americans the danger posed by Israel's conquest of the Sinai to détente. On the contrary, the more successes garnered by the IDF at the operational level, the more the Soviet position escalated and the threat to the relations between the superpowers was exacerbated; the United States was moved to rein Israel in, and the Egyptian grand strategy proved itself. Ironically, the IDF's successes at the operational level played into the hands of the Egyptian grand strategy and helped to fulfill it. In this context, neither the Israeli government nor the IDF ever attained relevance or effectiveness in foil-

## THE PICTURE BECOMES MULTIDIMENSIONAL

ing Egypt's grand strategy. Many years would pass after the end of the war before Israel even understood what had happened in that war in the wider sense.

One of the additional elements missing from an Israeli military decision was insufficient destruction of the enemy military (to a large extent as the result of American intervention). The fact that the 3rd Army was besieged but not destroyed, that the erosion of the 2nd Army was limited, and that the Egyptian military as a whole did not collapse but continued to maintain tactical opposition capability and residual defiance capability made it possible to claim that at the field levels too, the war ended in some sort of stalemate rather than a decision. This experience demonstrates once again the importance of destroying the enemy force as another essential component of decision. The reverse is also true: the Egyptians shaped a war in which the IDF's effectiveness was indirectly impaired, but the extent of direct destruction of the IDF's order of battle was limited. Therefore, the IDF was able to recover and adjust and ultimately achieve decision. Neutralizing the effectiveness of a military by attacking its plans and basic assumptions but without massive destruction of its forces enables it in certain cases to recover even while the war is still being fought, change its mode of operations, and thus avoid a defeat and even in the end achieve decision. Hence, even though this study stresses the critical importance of attacking the enemy's plans at every level of war, experience teaches us again and again that this is not enough, and the Clausewitz principle of annihilation must be upheld.

An important component of decision that Israel did fulfill was a demonstration of its tactical military superiority on the battlefield. Egypt had some impressive successes during the war, but those were attained through subterfuge and with the help of brilliant grand strategy and military strategy. By contrast, in most of the important and symmetrical engagements of force-on-force in the theater of operations, Israel clearly had the upper hand (particularly after October 14). This tactical superiority contributed to the consolidation of the military decision and even added to Israel's cumulative grand deterrence, and was among the elements driving Egypt away from the path of military conflict toward the path of diplomacy and peace.

Another central question is how this war offers an added perspective to the Clausewitzian understanding of center of gravity: clearly Egypt did not seek to attack the IDF's central mass, not even physical vulnerability nodes in the IDF system. Egypt sought to attack Israeli centers of gravity of an abstract nature, some military and others civilian–strategic (in fact, the centers of gravity at the grand strategy level). The military centers of gravity were Israel's defense doctrine, the IDF's war paradigm, the opportunity for the Israeli armored corps to maneuver and strike, the Israeli air force's freedom to fly, and Israel's reliance on its reserves. These centers of gravity were efficiently attacked, even if not by way of a direct confrontation but rather indirectly and asymmetrically.

Much more important for realizing the Egyptian war objective was attacking key civilian–strategic centers of gravity, including Israeli civilian stamina, Israel's self-confidence, the trust in and cohesion of the Israeli triad of government–military–civilians, the consensus that the 1967–73 status quo was a stable and desirable reality, and the Israeli-American relationship. This latter center of gravity included the American position on the Israeli presence in the Sinai and the American willingness (or lack thereof) to pay a steep price and take risks in the Eastern bloc-Western bloc system in order to ensure an Israeli victory. These centers of gravity were also attacked effectively. By contrast, the IDF focused on centers of gravity that were physical and operational in nature: the Egyptian mass of armor in the battle of October 14, the seam between the Egyptian fielded armies, the 3rd Army's supply lines, and the major roads leading westwards from the Suez Canal to Cairo.

It is true that the Egyptians took positions in the same theater of operations as the IDF, but they did so for a purpose completely different from the one the IDF had prepared for and intended. The IDF sought to fight the Egyptian military, whereas Egypt sought to attack Israel's defense concept and basic assumptions – at every level of the war – and to attack indirectly the country's civilian–political system. Despite this, both Israel and Egypt refrained from attacking one another's strategic home front by direct military means. Egypt had Scud and Kelt missiles, and the Israeli air force of course had the capability of attacking deep within Egyptian territory. Nonetheless, despite some relatively negligible footnotes, both sides limited themselves to using military means against military targets only.

It is worth asking how Israeli aerial attacks on Egyptian national infrastructures might have affected the progress of the war, especially in light of their centrality to the RMA/EBO concepts. The Israeli air force's capability of acting freely deep within Egypt was well known to the Egyptians: the Yom Kippur War broke out only three years after the end of the War of Attrition, which saw the air force launching massive strikes deep into Egyptian territory on refineries, electrical facilities, and other economic and military infrastructures. Therefore, one of the most important Egyptian decisions before the war was to cross the deterrence threshold, i.e., the decision about the willingness to risk deep strikes against national assets and to pay the price. In truth, the Egyptian decision not to be deterred by the threat of deep attacks was at the core of Egypt's pre-war situation assessments and the presidential directive to begin it.[41] Knowing that Egypt was prepared to pay a steep price for the war, one may ask how effective attacks against national infrastructures and other strategic assets might really have been. These would have had no direct effect on Egyptian military operational capability, and Egypt had already decided that they would not reduce its strategic freedom of action. Without a doubt a refinery is a valuable target, but when the enemy's war plans take its sacrifice into consideration, it is not clear to what extent attacking it in practice will affect the enemy. Not every valuable asset is a center of gravity, and not every long-range strike on a valuable asset is an effective strategic attack.

Thus, for example, attacking the national electricity grid is not in and of itself an attack on a strategic center of gravity. The government's inability to provide its citizens with electricity, contrary to their expectations, reveals the full extent of its weakness and may, within a sufficient period of time, upset the abstract strategic center of gravity – the cohesion and trust between civilians and their regime (i.e., the government's inability to meet civilians' expectations to provide a regular supply of electricity would bring about the civilians' lack of trust in their government and lead to their withdrawing support for the government's war effort). Still, this analysis is relevant primarily to democratic systems. The picture is more complex when the attempt is to challenge the stamina of a third world dictatorship. There are third world dictatorships that are based not on trust and cohesion between the regime and the civilians but on coercion, and the regime may well be willing to sustain certain blows to its home front. The will to fight is less dependent on the citizenry and much more on the stewards of state, and those have already spoken.

Furthermore, in the battle of rationales between Israel and Egypt, Israel sought to fight a war that would be decided on the basis of military effectiveness, whereas Egypt sought to shape a war that would hinge on a test of the sides' stamina and lasting power (table 3.3). If so, had Israel produced a main force application thread of attacking valuable assets that had no direct effect on Egypt's fighting capabilities or strategic freedom of action, it would actually have been operating on the basis of a rationale of testing the sides' stamina, namely in the service of the Egyptian rationale. A country going to war must decide what its main force application thread will be, e.g., attacking the enemy's armed forces, or denying it strategic freedom of action, or attempting to influence the will of the enemy's leadership, or attacking the enemy's fighting resources, and so on. Israel, which prefers short wars testing the sides' military effectiveness, must focus on attacking its enemy's military capability or its strategic freedom of action. It must not engage in wars where it tries over weeks or months to affect the enemy's will or resources, or in wars that test the stamina and staying power of both sides, a contest as to which side will outlast the other.

Historically, the argument that the enemy's political will is affected by attacking national resources and exacting a high economic price fails to hold water, at least in periods of times measured in hours or days. The German missile and aerial attacks against British cities and attacks on German cities by Britain and the United States in World War II, as well as

**Table 3.3** Asymmetry in Israeli and Egyptian War Doctrine, Yom Kippur War

| | Israel | Egypt |
|---|---|---|
| Victory (namely, attaining a political objective) through military decision | Decision on the operational level did not lead to victory (political) | Victory (political) even without a military decision |
| Decision defined as impairing the enemy's capability to act effectively | Yes – operational capability | No (yes, through a change in the nature of the war though not through destruction of force) |
| Decision achieved by attacking the enemy's operational centers of gravity | Yes | No; the important center of gravity was strategic/conceptual |
| Desiring a prime major battle | Yes | No (until the misjudged attack of October 14) |
| Engaging in the same theater of operations | Yes | Yes |
| Engaging in the theater of operations for the same objective | Yes (Israel's original intent, in practice – no) | No |
| Strike capability/stamina | Strike capability | Stamina (until the misjudged attack on October 14) |

the American attacks on North Vietnamese infrastructures in operations such as Rolling Thunder and Linebacker II, did not prove that exacting a price in and of itself undermines the political will to fight. Even the air campaign in Kosovo did not produce clear results in terms of hours or days, and it is not clear to what extent Kosovo's fate was sealed by the air strikes and to what extent by the threat of a ground invasion and the change in Russia's political stance.[42]

Therefore, it is crucial to distinguish between deep air strikes affecting the military progress of the war directly in a timetable measured in hours and days, such as an attack affecting military operational capability or directly narrowing strategic freedom of action on the one hand, and on the other hand, attacks whose significance lies in exacting a price and are meant to affect profit-and-loss considerations and the objectives of the enemy's political echelon in a timetable of weeks and months. Moreover, experience teaches that attacks exacting an economic price perhaps affect the period after the war – they extend the post-war reconstruction period, raise its cost, and possibly create some sort of deterrence – but they barely affect the course of the war in the time constants relevant to a decision campaign (as opposed to a campaign of attrition).

## The War of Attrition: Did Israel really Achieve Decision?

The War of Attrition (1969–70) saw the early though not yet fully developed signs of the war rationale Egypt employed in the Yom Kippur War, but it was implemented in a much more limited mode. The Israeli center of gravity that Egypt sought to attack was Israel's military and civilian stamina (namely, the civilians' will to fight and their support of the government in its bid to continue fighting). Thus already Egypt sought to divert the war from a test of the sides' military effectiveness to a test of stamina, demonstrating to Israel that the price of remaining in the Sinai would exact a heavy cost in Israel's most precious currency, i.e. casualties, and in disrupting the economy for extended periods of time.

Egypt's military mode of operation consisted of artillery fire and raids on the IDF's front lines deployed in the operational space along the Suez Canal, but not for the sake of undermining the IDF's operational capability or crushing its original structure, but in order to force Israel to maintain a large military presence made up of both regular military units and reserve units on the shores of the canal for extended periods of time. This was a source of pressure on the economy and society, and gradually, over time, Israel's will to fight was "supposed to" erode. This erosion, it was hoped, would gain momentum with Israeli soldiers regularly killed in the course of daily, routine activities, rather than in the context of large battles with obvious benefits.

Egypt operated on the basis of two assumptions: first, that Israel would not escalate the War of Attrition to a comprehensive war and would not cross the Suez Canal in order to achieve decision against the Egyptian military; and, second, that Israel was not capable of retaliating against the Egyptians in kind. As a non-democratic, non-industrialized country lacking a free press, a free economy, welfare services, and a significant and assertive middle class, Egypt was seen as having an advantage in national stamina and staying power, allowing it to outlast Israel's free market democracy. In fact, Egypt was willing to sacrifice its three cities near the canal, turn millions of Egyptians into refugees, sustain tactical military blows from IDF fire, and even sustain severe damage to refineries, dams, and other state infrastructures. Still, not every national asset, precious as it might be, is a center of gravity. In order to inflict a blow to a center of gravity linked to national stamina, it is necessary to identify a clear and significant weakness in the political–civilian fabric and relevant to the given context. It is unclear what effect Israel's strategic attacks had on the Egyptian political–civilian fabric; even if they did have an impact, it could not be measured in hours or days, but perhaps cumulatively over the course of many months.

Nonetheless, the common claim is that Israel's air strikes on national infrastructures deep within Egypt broke the Egyptian will to fight and prompted Egypt to request a ceasefire. That is, the deep air strikes undermined Egypt's basic assumptions, and it was actually Egypt's stamina that gave out first. There are those who claim that it was not the physical effect of the attacks themselves that motivated the Egyptians to seek a diplomatic exit from the war, rather that these attacks exposed the weakness of the Egyptian leadership, which was unable to provide appropriate defense to the country's strategic rear. It was an attack on the prestige and legitimacy of the leadership (and this, apparently, was the real Egyptian center of gravity). At best, Israel succeeded therefore in exerting pressure on Egyptian national infrastructures, thus revealing the weakness of the Egyptian leadership to the Egyptian people and limiting Egypt's strategic freedom of action to continue fighting, and prompting it to seek an exit from the war despite its capability of continuing to act effectively against Israel at the operational level.

However, the opposing view is also tenable: not even a few hours had passed after the ceasefire before Egypt forwarded SAM batteries to the Suez Canal, a blatant violation of one of the most essential terms of the ceasefire agreement. That would seem to indicate that Egypt was willing to continue fighting if Israel responded to the Egyptian violation. It was actually Israel and the United States, both weary of the conflict, that were reluctant to resume fighting and therefore decided to ignore the Egyptian violations. Moreover, the end of the War of Attrition did not lead Egypt to despair of the military option. On the contrary: immediately upon the end of the War of Attrition in 1970, Sadat began to prepare for a comprehensive war. It was originally scheduled to begin in 1971, and was postponed several times until 1973. That would seem to indicate that the national infrastructures did not represent so critical an asset that attacking them would generate a strategic reversal on Egypt's part or cause it to recoil from war.

In any event, the War of Attrition can be summarized as follows: Egypt sought to achieve the war's objectives even without a military decision against Israel, whereas Israel, if we posit that Israel did in fact defeat Egypt, did so without inflicting a blow to the Egyptian military's original structure or its operational capability (though Israel did demonstrate that the Egyptian military was inherently incapable of foiling the air force's deep strikes, i.e., it was not capable of acting effectively in this context). Neither side acted against its rival's military's operational centers of gravity and no prime major battle was fought. Although the Egyptians did assume positions in the battlefield, they did so for the purpose of an indirect attack on Israel's civilian–political–strategic centers of gravity.

## Who Won in 1956?

Intuition suggests that the Sinai Campaign (1956)[43] ought to be categorized as a "simple" symmetrical war, because the buildup of the forces of both sides was similar and both sides viewed the war fundamentally through the same prism: a clash between regular field masses. Nonetheless, the Sinai Campaign is included in this chapter on asymmetrical, multilevel, complex wars for two reasons. First, the two sides took up positions in the theater of operations for fundamentally different strategic objectives. Second, the story of the war is much more complex than the "simple" symmetrical clash between masses in the theater of operations. The story of the war is fundamentally different from one level to another; the outcome of the war is different from one level to another; and the correlation of the outcome from one level to another is either very limited or nonexistent. Particularly blatant is the lack of a cause and effect relationship between the military end state and the political end state. The bottom line is that the war did not test the military effectiveness of the sides but rather their ability to leverage the international community.

The main objectives of the war common to France, Britain, and Israel was toppling the Nasser regime and putting a nonmilitant, pro-Western leadership in its place. France and Britain also wanted to secure their national interests in the Suez Canal and to strengthen their declining status as world powers. Israel sought to attain certain national objectives of its own: open the Straits of Tiran to shipping, bring an end to fedayeen terrorism, prevent a shift in the regional balance of power as a result of the Egyptian–Czech arms deal, and strengthen its deterrence capability. Moreover France, deep in its war against the Algerian FLN and lacking good military access to the arena, needed Israel's help in attacking Egypt, while Israel, wanting nuclear and advanced weapons assistance from France, wanted to cement its alliance with France.

At the grand strategy level, Israel assumed that siding with Britain and France afforded it the military and diplomatic capability to realize the objectives of the war, and it also assumed that Britain and France would gain the support of the United States. By contrast, Nasser's reliance on the Soviet Union was not self-evident at that time. The ties between Cairo and Moscow were not yet close, and many in Egypt thought that Egypt should align itself with the United States (Nasser himself claimed that nationalizing the Suez Canal in order to finance the construction of the Aswan Dam was a less anti-American step than accepting the financing package offered by the Soviets). In those days, the Soviet Union's attention was focused on suppressing the Hungarian uprising and on fears of a similar uprising in Poland. In the Kremlin, there were voices calling to create an equation whereby the big powers would be allowed to intervene with force in order to protect their critical interests (the Soviet Union in Hungary, Britain and France in Suez). In the end, however, the Soviet Union decided that it was likely to strengthen its status in the Arab world by helping Egypt, and that the threat of an East-West escalation would lead the United States to dissociate itself from Britain and France and undermine the unity of the Western bloc. In fact, the Soviet Union applied diplomatic pressure and even veiled nuclear threats (the Nikolai Bulganin missives), persuading President Eisenhower that there was a clear and present danger of an Eastern-Western bloc conflict or, at the very least, a chance of Soviet military penetration of the Middle East. This concern led to an extreme and uncompromising joint stance of the US and the Soviet Union whereby Britain, France, and Israel had to withdraw at once from Egyptian territory. The three allies had no choice but to comply. Israel's basic assumption that joining France and Britain would ensure it success in enlisting international dynamics collapsed.

The fact that in the political end state not a single French, British, or Israeli soldier was left on Egyptian territory, that the international community acquiesced to Egyptian control of the canal, and that Nasser succeeded in saving his armed forces from destruction enabled Nasser to claim – with a great deal of justification – that he won the war. The primary objective of Israel, France, and Britain was not achieved, whereas Nasser was left defiant and in control. The fact that he faced the France-Britain-Israel coalition and remained standing strengthened his status as a leader in the Arab world and his claim to leadership of the non-aligned nations. This was a preview of the "winning by not losing" phenomenon. However, Israel did attain some indirect achievements of a wider context, in particular a boost to its self-confidence in terms of its military capabilities and a change in the way the Arab world and even the global powers saw Israel. Above all, the war initiated a shift in the United States' perception of Israel, and was an important milestone in creating the strategic link between the two nations. The war also made Israel a valuable ally of the non-Arab regional powers (Iran and Turkey) and increased Israel's influence in Africa. The war thus resulted in a better fundamental reality for Israel (and Israel also gained international assurances about freedom of shipping, demilitarization of the Sinai, and an end of fedayeen activity, though these did not hold up when put to the test).

At the military strategy level, the story of the war is very different. Israel sought to optimize some contradictory considerations. First, it wanted to provide the friendly powers a pretext for war by taking a step that would be seen as a direct threat against the canal zone. Second, the lack of trust in Britain and France made Israel unwilling to commit itself to a comprehensive war before the global powers' intervention was a fait accompli. Israel preferred to postpone the main clash with Egypt until after the powers' invasion, when the Egyptian military would be unable to reinforce

## THE PICTURE BECOMES MULTIDIMENSIONAL

itself or to launch counterattacks. Only third did Israel want to realize its own particular war objectives directly and through independent military action. Israel sought to attain a quick decision against the Egyptian armed forces, as it was concerned that another front would be opened or that Iraqi forces would enter Jordan. In fact, Israel fully succeeded in realizing its plans at the military strategy level.

At the military strategy level, there was clear asymmetry between the sides. Egypt was concerned that nationalizing the Suez Canal would lead to a French–British invasion. Therefore, even before the war, Egypt began to move forces from the Sinai front to the Cairo and Suez Canal regions, and when the British-French attack began, the Egyptian force received an order to withdraw from Sinai. Israel attacked a military system that was in the process of being thinned out, and that started to withdraw immediately at the outbreak of the war and before the prime major battle.

As to the field moves themselves: from the perspective of the higher levels of war, Israel's parachuting into the Mitla Pass was meant to provide Britain and France with the necessary pretext for entering the war. Penetrating to the center of Sinai was meant to enable joining the parachuted force (or extricating it, in case the powers failed to intervene). The early breakthrough to the Abu-Ageila zone before the global powers attacked Egypt was an operational mishap (resulting from insufficient clarification of the strategic concept to the fielded echelons). The main effort was to attack the rear of the Egyptian force that was already withdrawing from Sinai. From the perspective of the operational level, Israel created an almost simultaneous breakthrough of the Egyptian defensive lines into the depths of the Sinai Peninsula, and thereby denied the Egyptian military its capability to fulfill its designated function. The Egyptian formation, which was already fragmented and was not continuous, was pierced and penetrated in its entire depth, and its operational level posture was undermined. This was done (purposely) with limited friction between the armed forces, through bypassing the centers of mass and with relatively few large battles.

Undermining the Egyptian operational level posture was achieved by attacking some physical centers of gravity. The center of gravity anchoring Israel's hold on the western Sinai was the Mitla Pass, into which a force was parachuted at the outbreak of fighting. The center of gravity of the second Egyptian defensive line, which was also the major center of gravity of the entire campaign and where the Sinai Campaign was in fact decided, was the Abu-Ageila-Jabel Livni strong point, a major junction in the center of the Sinai Desert with roads towards the north and the south of the peninsula. The Egyptians had constructed fortifications in this zone, and to its rear they had placed their main mobile armored reserves. Another center of gravity was the Rafah strong point, located on the northern road, the shortest between Israel and the Suez Canal. The Egyptians had constructed fortifications in this location as well. In fact, attacking these

## THE COMPLEX ASYMMETRICAL WAR AGAINST A REGULAR OPPONENT

Egyptian compounds, together with parachuting into the Mitla Pass and breaking through the Egyptians' eastern defensive line near the border (Um Katef-Kusseima) led to both the Egyptian defensive system's loss of capability to fulfill its designated function throughout the entire depth of the Sinai and to the Egyptian loss of the will to fight, causing the collapse of the Egyptian force in the Sinai and decision at the operational level.

Nonetheless, most of the Egyptian force succeeded in fleeing the theater, and most of Egypt's forces survived the campaign. The very fact that Egypt had any forces at all left after the war enabled the Egyptians to claim that they had not been defeated, despite their military's inability to fulfill its designated function in the theater of operations. This fact yet again demonstrates the necessity of the "simple" destruction of the enemy's forces as an additional component of military decision, and in one's ability to translate field success into a victory.

From the perspective of the lower levels of the war, the story of the war is again very different and is characterized by mediocrity, if not worse. Most of the Israeli fighting units comprised reserve forces that suffered from a flawed mobilization procedure, lack of training, low fighting capability, and severe maintenance and logistical problems. Equipment failed to arrive from the depots to the fielded units, commanders were unfamiliar with the plans, and field intelligence was far from perfect. The regular forces also suffered at times from tactical disappointments, and even the elite paratroopers dropped into the Mitla became embroiled in a difficult, costly battle.

However, these tactical and logistical weaknesses were isolated at the

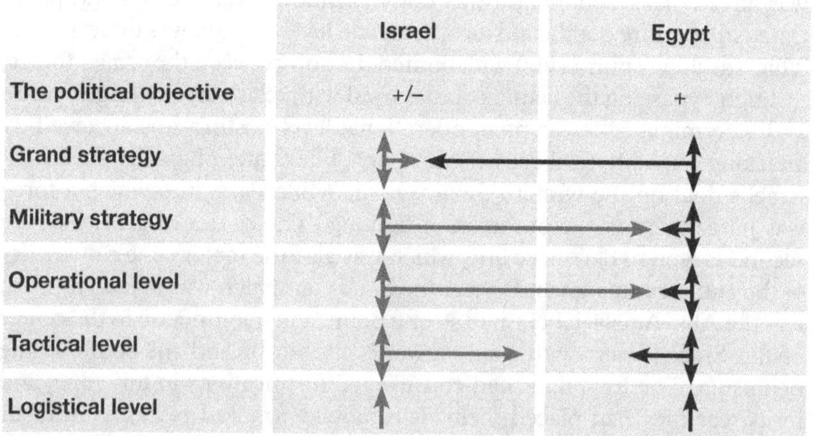

|  | Israel | Egypt |
|---|---|---|
| The political objective | +/− | + |
| Grand strategy | | |
| Military strategy | | |
| Operational level | | |
| Tactical level | | |
| Logistical level | | |

**Figure 3.2** The Sinai Campaign: Different outcomes at the different levels of war

lower levels, and did not hamper the IDF's capability of fulfilling its plans and rationale at the operational and military strategy levels. Despite the tactical and logistical weakness, Israel succeeded in providing the friendly powers with a pretext for the war, piercing the Egyptian lines and upsetting its equilibrium (particularly by maneuvering against terrain rather than against the enemy), and attacking the rear of the Egyptian forces retreating from the powers' invasion. By contrast, these successes were isolated at the higher military levels, and were not translated into an achievement at the grand strategy level or into a fulfillment of the political objective. The United States estimated that the Soviet threat was credible and therefore did not stand by the side of its natural allies, and so despite the attainment of the planned military end state, international circumstances did not enable the fulfillment of the desired political end state.

Figure 3.2 charts the various outcomes of the different levels of the 1956 war.

## Overview of Multilevel Asymmetry

It is thus possible to discern that in asymmetrical wars between regular state opponents, there is a tendency to go beyond the mutual desire to fulfill the war's objectives via a direct confrontation between the forces in the theater of operations for the purpose of achieving a military decision. This is no longer a "simple" clash between military masses seeking to destroy one another on the battlefield, but rather multileveled and multidimensional wars. Armed forces increasingly take positions on the battlefield to serve complex rationales, not necessarily tactical only but also on the operational, strategy, or grand strategy levels. Despite the self-evident interrelationships between the levels of war, it is also possible to discern autonomous confrontations at every level; often, the correlation between outcomes on lower and higher levels is limited to non-existent. Moreover, in some cases, the political end state is not constructed on the basis of the military end state but in disregard of it. The political end state may even start to form before the military outcome is clear, or even before the outbreak of the war. The military actions might provide only the context or the catalyst to move towards the political end state, but will not create it directly. Therefore, if in a classical war it was necessary to wait until the military end state was clear and only then to turn to shaping the political–strategic end state, in complex wars it is sometimes possible to formulate the political end state before the military outcome is clear or even before the war has begun.

Complex wars do not necessarily test only the sides' military effectiveness, but rather a series of additional issues, such as stamina, resources, and enlistment of the international community. In many cases, there is a war going on in the background over the nature of the war: each side strives to force a war that tests a different issue and plays to its own relative strength and to the opponent's relative weakness.

In the case of asymmetry of national stamina, it is sometimes possible to discern a pattern in which one side tries to shape a war in which stamina is tested, while the other side makes an effort to force a war testing military field effectiveness. The side trying to place stamina at the center of the war seeks to create a situation that is on the one hand intolerable and exhausting to the other side over time, and on the other hand denies it an opportunity for decision inter alia by diluting operational centers of gravity or by not taking positions for prime major battles.

In those asymmetrical, multi-issue wars, decision is not derived only from the extent to which the enemy's armed forces are destroyed, but from the question of which war was fought: the one we prepared for, or the one the enemy prepared for. The side that can impose the type of war for which it has prepared can act effectively to fulfill its designated objectives, whereas the other side is rendered less relevant already from the outset. In other words, the other side's capabilities and plans will be ineffective in a particular context, and hence it is almost as if the first side has reached decision. Moreover, the enemy's capability to continue operating effectively cannot be measured only in physical, tactical, or even systemic-operational terms, but must also be measured by the capability of the military force to fulfill its designated function at the strategy and grand strategy levels. Nonetheless, this does not negate the need to annihilate the enemy's military forces: this "simple" destruction remains an essential constituent of decision.

The center of gravity attacked is not just the place where the enemy's military mass is densest, but often belongs to the strategy or grand strategy level, whether it bears physical characteristics or more abstract ones, such as trust and cohesion of the government–military–civilian triangle, the willingness of the political and public systems to continue supporting the war objectives, war plans, basic assumptions about defense and defense conceptions, defeating the enemy's rationale or paradigm, undermining its operational level posture, or relationships with allies. When attacking these complex centers of gravity, it is sometimes possible to realize the objectives of the war even if the enemy's force can continue to act with some measure of effectiveness or another at the field levels.

However, not every valuable asset is a center of gravity, and its categorization depends on the nature of the war and on the force application thread (such as inflicting a blow to the enemy's fighting capability, to its strategic freedom of action, to the will of its leadership, to its war supporting resources, and so on). In a war seeking a quick test of the sides' military strike capability, a center of gravity can be recognized by the fact that attacking it would have a direct and immediate effect on the military

## THE PICTURE BECOMES MULTIDIMENSIONAL

progress of the war: a blow to the enemy's fighting capability or denying the enemy its freedom of action to continue fighting according to the paradigm it desires. However, even in extended wars of attrition that seek to test the stamina and staying power of the two sides, in which each side attempts to outlast the other, not every valuable asset is a center of gravity. Striking at assets is meant to chisel weakness in the national fabric over time. In a third world dictatorship, that might be the leadership's will to fight, whereas in a Western democracy it is usually public support for the continuation of the war in the face of its cost. Often, in asymmetrical tests of stamina, democracies are more vulnerable.

# 4

# Asymmetrical Wars against Non-State Opponents

## Same Theater of Operations, Different Objectives

This chapter describes and analyzes wars between states and non-state enemies that demonstrate the following:

1. The non-state enemy does usually not adhere to classical definitions: it does not seek decision against a regular armed force, does not seek to inflict blows on operational centers of gravity, and does not take positions for a prime major battle.
2. Centers of gravity of interest to the non-state enemy are the stamina of the civilian–political system of the state enemy, in order to undermine its will to fight.
3. Undermining the political–civilian will to fight is achieved incrementally over time, thereby necessitating prolonged fighting.
4. Undermining the political–civilian will to fight is achieved by creating a situation that is on the one hand intolerable over time (usually by inflicting ongoing blows to the opponent state's expeditionary forces), and on the other hand, denies the state force the opportunity to achieve decision (disappearing, avoiding large battles).
5. In order to maintain this situation, the non-state enemy does not need to preserve full fighting capability, but only residual defiance capability.
6. The benefit the non-state enemy derives from the engagement on the battlefield is not measured in tactical-physical terms, and therefore it can fulfill its designated strategic–political function even if it absorbs significant tactical blows.
7. The political end state of such an asymmetrical war does not emerge directly from the military outcome but from much broader contexts. The connection between the war's outcome at the military levels, particularly at the field levels, and the strategic–political outcome is not direct.

## The Westphalian Vietcong versus the anti-Westphalian Jihad

It is obvious that the irregular or non-state enemy represents a certain type of asymmetry, with its own world of unique analytical tools and terminology. This type of enemy is not new, and the world has seen many examples of war against such entities, e.g., Napoleon on the Iberian Peninsula (1808–14), Britain in the Boer War (1899–1902), and Germany in World War II, inter alia in Yugoslavia and Russia. In the post-World War II era there have also been many campaigns against irregular enemies, notably Britain's campaigns in Malaya (1948–60), Aden (1963–67), Kenya (1952–59), Cyprus (1955–59), and Greece (1946–49); the French campaigns in Indochina (1946–54) and Algeria (1954–62); the Soviet campaign in Afghanistan (1979–89) and the two Russian campaigns in Chechnya (1994–96, 1999–2009); the Dutch campaign in the East Indies (1945–49); and of course, the United States in Vietnam (1964–75), and more recently in Afghanistan (2001–) and Iraq (2003–). Clausewitz himself learned about guerilla warfare – the *kleinkrieg* (small war) and *volkskrieg* (folk war) – primarily through Napoleon's experience, and that unusual experience led him to reexamine and qualify some of his earlier opinions.

It is important to distinguish between at least two different types of non-state opponents: those who at the end of their struggle seek to be part of the international state system, and those whose final objective is to undermine it. The Viet Minh, the Vietcong, and the Chinese communists engaged in a non-state, irregular struggle as an interim stage until the attainment of the desired end state in which a sovereign nation state would be established as part of the existing international political system. Therefore, despite being in a non-state stage, they acted according to policies that (from their point of view) were pragmatic, rational, and utilitarian; they considered profits versus losses; and finally, they became partners to a dialogue consistent with basic state rationale. By contrast, after the March 2004 terrorist attacks in Madrid, al-Qaeda released a statement according to which "the international system established by the West ever since the Westphalia Accords is destined to collapse, and a new international order will arise under the leadership of the glorious Islamic nation." The Peace of Westphalia in 1648 put an end to the European wars of religion, and was considered a defining event accelerating the maturation of Western civilization from a hierarchy of rulers and mainly religious loyalties and motivations to a "Westphalian system" – the international system familiar to us to this day based on nation states, the concept of sovereignty, and pragmatic-secular diplomacy.

Indeed, a confrontation with an opponent that avoids state-like rationale not only as an interim stage and as the result of circumstance but

## SAME THEATER OF OPERATIONS, DIFFERENT OBJECTIVES

also out of permanent ideological, cultural, and strategic choice generates complex and unfamiliar difficulties. By its very nature such an opponent is liable to be formless and incoherent. It might not necessarily have a central decision maker, or structured decision making processes or profit-and-loss considerations. It does not necessarily operate to fulfill realistic, attainable objectives, and it is not bound by state-like timetables. Often, the ideology is unclear, and its followers are believers whose numbers and mutual connections are undefined. Moreover, some of these opponents might view the phenomenon of war in a fundamentally different way from the way we do: we view war as a necessary evil, sanctify the value of human life, examine every action on the basis of its outcome, and conduct our affairs by characterizing the problem and solving it. Other cultures, such as jihad-driven, view war as a preferred and lofty cultural choice and as an historical religious imperative free from cost-benefit considerations and practical timetables. There are people among them who extol death and the act of fighting – not its outcome – as that which really matters.

This memorandum does not have war against an irregular or non-state jihadist enemy as its principal subject, rather examines a few examples to demonstrate the similarities and differences with regard to classical military doctrine. However, each of the cases mentioned above is characterized by particular circumstances, and it is incorrect to categorize them uniformly. Indeed, along with some stinging failures – such as those of the US in Vietnam and the French in Algeria – Britain, for example, succeeded unequivocally in its Malaya campaign and attained the political–strategic end state it sought.[44] Thus, contemporary Western democracies can achieve military decision against the guerilla enemy.

To what extent does war against a non-state enemy conform to the basic assumptions of classical military doctrine? The answers to this question vary, and derive in large measure from the nature of the regular state's center of gravity that the non-state opponent aims to attack, and from the pattern of action of the non-state side.

More commonly, the non-state opponent seeks to avoid a direct confrontation with the regular forces in a prime major battle, understanding that its chances of rendering a blow against the regular force's operational center of gravity and crushing its original structure are low. In other words, a non-state force will almost never achieve a military-operational decision against a regular force. Therefore, the non-state side will usually choose to inflict blows to the enemy's military or civilian targets in order to indirectly undermine the civilian–strategic centers of gravity of the regular state opponent. The non-state side is also likely to use its own civilians as human shields, on the assumption that unavoidable harm to them caused by the regular forces would shock the regular forces' civilian home front and drive a wedge between the state's military and its civilian–political rear. In other words, the non-state opponent prefers to contend with the political–civilian stamina of its state enemy, rather than with its military strike capability. It seeks to attain victory (to realize the political objectives of the war) even without a classical military decision. Nonetheless, from the outset, it tries to build up its force and to shape a war of a nature that will make it difficult for a regular armed force to act effectively against the non-state entity.

## The Tet Offensive: American Tactical Success, Vietnamese Strategic Success

The Vietcong did not intend to expel the Americans from Vietnam through a military decision against the American armed forces. It killed American soldiers in the jungles of Vietnam (and even dragged the Americans into situations in which Vietnamese civilians suffered casualties) in order to puncture the American television viewers' and voters' will to support the fight, thereby generating American internal political processes that would result in a US withdrawal from Vietnam. By contrast, the objective of the American armed forces was more focused on (at least during certain stages of the war) dealing direct physical blows to the Vietcong's fighting capability through search and destroy operations and attempts to create relatively large scale battles. While both sides assumed positions in the same theater of operations, they did not do so for the same objectives: the Americans wanted to inflict a blow against the Vietcong and their resources mainly in order to erode the organization's military capability, while the Vietcong sought to cause American casualties in order to undermine the civilian home front in the United States. From the American perspective, the direct outcome of battles against the Vietcong (the so-called "body count") was more important, whereas the Vietcong was almost indifferent to direct tactical variables such as the number of casualties sustained or the question of which side was temporarily in control of a certain area where the specific battle was conducted. It assumed positions in the theater of operations mainly in order to affect the strategic–political level of the war – the cumulative and incremental erosion of American determination and consciousness, a result of the long series of battles and the lack of any clear achievement or offer of a military exit from the war.

This point is demonstrated, for example, by the consequences of the asymmetrical ramifications of the Tet Offensive.[45] In the January 1968 Tet Offensive, the Vietcong operated extensively with large formations and in a large number of places, and for the first time it also included the use of large regular North Vietnamese forces with high signature characteristics. Large battles were fought in many locations, some of which resembled the urban battles of World War II. Therefore, the Tet Offensive, at the tactical-operational level, played into the hands of the Americans, and in the theater of operations, it was an unequivocal failure from which neither the Vietcong nor the North Vietnamese military recovered until the middle of 1971. (The Vietcong alone lost some 45,000 fighters in this offensive.) In the Tet Offensive, the two sides assumed positions with large high signature forces in the same theater of operations, and in tactical-operational terms of destroying the enemy's mass, it yielded clear American success.

Nonetheless, at the strategy level and particularly the grand strategy level, the Tet Offensive marks the watershed when the American public and media lost faith in the administration and the armed forces. Until then, the American public believed that it was possible to win the war and that their political and military leadership knew what it was doing. This faith was undermined when the news broadcasts presented the simultaneous enemy offensives in dozens of South Vietnamese cities, among them three days of fighting in the capital of the south, Saigon, including attacks on the American embassy and the headquarters of General William Westmoreland, commander of the American forces in Vietnam. A famous anecdote recounts President Lyndon Johnson's response to the series of gloomy reports filed by noted television reporter Walter Cronkite, who was in Vietnam in early 1968: "If I've lost Cronkite, I've lost middle America." The benefits each side garnered from the Tet Offensive were on completely different levels and axes. Thus in April 1975, after the end of the war, Colonel Harry Summers bumped into a North Vietnamese colonel named Tu. The American claimed that the North Vietnamese never won against the Americans on the battlefield. "That may be so. But it is also irrelevant," was the North Vietnamese response.[46]

And indeed, despite American successes in every tactical battle worthy of mention in the Vietnam War, the United States did not achieve a military decision against the Vietcong. It did not succeed in translating its field achievements into higher levels of the war. It also failed to shape a war that would hinge on a test of the sides' military effectiveness or resources, where the United States had a relative advantage. Clearly, the Vietcong did not achieve a military decision against the United States armed forces either, but the Vietcong did succeed in wearing down the American public's stamina and the trust and cohesion of the government–military–civilian triad, to the point that the US withdrew from Vietnam, even though the original structure of its military was not impaired and it could have continued to operate militarily against the Vietcong. In order to erode the American civilian–strategic will to fight, the Vietcong did not have to show full classical military effectiveness (which in fact suffered a severe setback as a result of field engagements), and merely had to preserve its residual defiance capability to enable it to continue to maintain a military situation that on the one hand over time was intolerable to the American public and political system, and on the other hand, had no military exit (i.e., denied the American armed forces the opportunity for a military decision). The effectiveness of the Vietcong as a fighting force cannot be measured in tactical or physical terms – i.e., in field achievements against United States troops – rather in terms of the strategic and grand strategic effects the Vietcong managed to generate. And, since the Vietcong was capable of generating the same effect on the American political–civilian system as long as it continued to maintain residual defiance capability, the question of what constituted a military decision against the Vietnamese became complicated. Clearly, the Vietcong would have been directly overpowered (namely, it would have been incapable of operating effectively to realize its designated function at the strategy and grand strategy levels) only if it had been unable to undermine further the American political–civilian will to fight.

Examples of modern wars against non-state opponents that conform to the basic assumptions of classical military doctrine in which the two sides assumed positions in the same theater of operations and for the same purpose are rarer but do exist as specific stages of campaigns against irregular opponents or in specific battles. One prominent example is the Dien Bien Phu campaign (1953–54) between French expeditionary forces to Indochina and the indigenous Viet Minh. The political objective of both sides was to improve their bargaining position at the forthcoming Geneva conference, and both sides sought to do so through a clear military success and even a local military decision.

Not only the regular French forces, but also the non-state Viet Minh tried to attain a military decision by attacking the opponent's operational center of gravity in a prime major battle. The French chose to advance a force of 15,000 soldiers onto open ground, in order to lure the Viet Minh into taking position for a prime major battle, in which clearly the advantage was supposed to have been held by the French regular forces. Contrary to the usual rationale and pattern of action of guerilla organizations, the Viet Minh accepted the challenge and assumed the ambitious task of attacking the center of the military mass of the French expeditionary forces in the said area and destroying it (not just generating cracks in France's political will to fight by causing casualties). And indeed, the Viet Minh moved some 50,000 fighters and artillery and anti-aircraft assets into the Dien Bien Phu sector, and took positions for an extended set-piece battle with high intensity (209 days, including a 54-day siege), in which over 2,000 French soldiers were killed and the rest captured. Thus, the Viet Minh attained a local military decision – and even more.

## The First Lebanon War: Surprisingly, Maneuver Achieves Military and Strategic Decision

One of the less typical examples of a war between a regular military and a non-state opponent in which the regular side still succeeded in conducting

## SAME THEATER OF OPERATIONS, DIFFERENT OBJECTIVES

a war more or less according to classical military doctrine and achieving a clear decision is the IDF campaign against the PLO in the First Lebanon War (1982). The PLO's operational center of gravity was its deployment in southern Lebanon near the Israeli border, which enabled it to attack the Israeli home front with rocket fire and through infiltration of squads of terrorists. The PLO's strategic center of gravity was the opportunity to establish a state-within-a-state in the territory of Lebanon, a dysfunctional state forced to host the PLO after the PLO was previously expelled from Jordan when it attempted to set up a state-within-a-state. An additional strategic center of gravity was the support the PLO received from Syrian military forces in Lebanon, which were also attacked by the IDF.

The key to decision over the PLO in 1982 was the fact that in face of Israeli maneuvers in southern Lebanon, the PLO, unlike typical guerilla organizations, did not scatter or hide its fighters and did not melt into the natural or civilian background, rather withdrew its forces northwards while conducting itself in large, semi-military formations with a relatively high signature. Perhaps the Palestinian PLO members differed nationally and ethnically from the local southern Lebanese population (except in the Palestinian refugee camps), or there was an attempt to advance from the guerilla stage to that of regular warfare or an attempt to protect the refugee camps. Whatever the reason, the said pattern of the PLO's reaction to Israeli maneuvers meant that the IDF was not denied the opportunity for decision, and the maneuver managed to force the PLO out of southern Lebanon and into Beirut within one week. The subsequent two-month siege of Beirut brought about the PLO's final expulsion from Lebanon to Tunis (far from Israel), dealing an absolute and final operational and strategic defeat. As a political entity, of course, the PLO has survived to this day, but it is no longer the same organization with the semi-military operational capabilities that controlled Fatahland in the 1970s.

As will also be discussed below, the differences between the attributes, nature, rationale, and pattern of action of the PLO in 1982 and those of Hizbollah in 2006 produced – in the same arena and in both cases against a non-state opponent – completely different wars. Hizbollah is a hybrid organization, in part something like a division of the Iranian army, but in part also an authentic grassroots representative of the Lebanese Shiite population. As such, Hizbollah had the ability to blend in, something the PLO apparently lacked, and therefore Hizbollah was able to construct for itself two possible methods of operation: first, attacking the Israeli home front directly using rockets, with Israel responding with fire but without maneuvers; second, should Israel respond with maneuvers in southern Lebanon, staying there with a low signature and disappearing among the local population and conducting an ongoing guerilla war against the IDF with the purpose of striking at IDF soldiers, thereby indirectly striking at Israel's strategic center of gravity ("spider web" is the Hizbollah metaphor for Israel's lack of public-political determination to persevere in war). This suggests that while the PLO did not have a fitting response to Israeli maneuvers in southern Lebanon, for Hizbollah, such maneuvers afforded an opportunity for waging a guerilla war, and therefore represented a useful alternate method of operation (as far as Hizbollah was concerned). And while the PLO was a foreign element in Lebanon and its expulsion produced an absolute, final decision against it as a military force, Hizbollah, which is also an authentic indigenous organization emanating from the local Shiite population, can almost always come back and recoup its forces in Lebanon. Therefore, there is a real risk in the situation reverting to its original state at the end of any Israel maneuver against Hizbollah in southern Lebanon, and it is difficult for Israel to find a successful exit strategy for such maneuvers.

The particular features of the PLO's mode of action in 1982 enabled the campaign waged by Israel against it to come very close indeed to classical military doctrine: Israel's war objectives were achieved through military decision;[47] the military decision was marked by the absolute and irreversible damage done to the PLO's capability of acting against Israel effectively as a fighting and organized semi-military force; the military decision was achieved by attacking the PLO's operational center of gravity and even by attacking its strategic centers of gravity; and the centers of gravity were attacked in a prime major battle, or at least in a series of battles conducted as part of the same large, concentrated maneuver. The two sides took up positions in the same theater of operations and, certainly compared to wars between states and guerillas, used the same regular military rationale (i.e., the PLO basically did not operate according to the guerila rationale of disappearing among the population).

The lexicon for combat against guerillas usually does not fit classical military doctrine. When guerillas win (namely, attain their political objectives), their victories are not a direct result of a military decision, rather stem from a wide set of contexts and usually from an extended, incremental exhaustion of the state side's civilian–political will to fight. Guerillas seek to obscure their military-physical centers of gravity, and usually avoid the prime major battle. In order to realize their designated function at the higher levels of war, guerillas are not required to maintain full field effectiveness, and it is enough if they preserve their residual defiance capability, which will gradually exhaust the political–public consciousness of their state enemy and enable them to outlast the state opponent.

# 5

# Parallel War

## One War with Two Non-Convergent Campaigns

This chapter analyzes wars in which each side operates against its opponent in a different manner, without even necessarily taking positions in the same theater of operations. Each side attacks a different type of center of gravity, and every such campaign takes place in territorial or functional isolation from the parallel campaign being conducted by the opponent. The following points are discussed:

1. The doctrinal and theoretical background is expanded to encompass alternatives to Clausewitzian concepts, including attacking the enemy's plans and the strategy of indirect approach.
2. There is a tendency of cultures not subscribing to the Western tradition to expand the context of war to the "situation as a whole."
3. The "situation as a whole" approach is likely to purposely involve the civilians of both sides, to blur the distinctions between war and peace and between warfare and diplomacy, and is not limited to the boundaries of the theater of operations or defined periods.
4. Even in wars testing the military effectiveness of the sides, it is necessary to distinguish between:
   a. attacking an operational center of gravity to achieve a military decision, i.e., a critical and direct blow to the enemy's military operational capability, and
   b. attacking or threatening a strategic center of gravity to achieve a strategic surrender, i.e., denying the enemy's freedom of action to continue fighting.
5. The wars described in this chapter are mainly of the second type, i.e., the effective application of military force to deny the enemy's strategic freedom of action to fight or at least to attack the enemy's plans to wage a war based on a particular paradigm, and to impose a different paradigm, without directly attacking the enemy's fielded forces.

## The Tactical West versus the Strategic East

Basic Clausewitzian assumptions, described in detail at the beginning of this study, are not the exclusive approach to war, even among classical thinkers. Sun Tzu, for example, claimed that the primary method of achieving victory over the enemy was by "attacking the enemy's plans," i.e., attacking enemy strategy. In his view, the "ideal" war is one where the outcome is determined even before war erupts, and it is better to achieve a victory by putting the puzzle pieces of the winning strategy in place without ever resorting to the actual use of force.[48] The direct attack on the enemy forces is to be undertaken only when there is no better alternative, and it is only necessary to the extent it serves the comprehensive rationale of undermining the enemy's strategy. In effect, Sun Tzu places the responsibility for victory on strategy, and the tactical battle merely represents the inevitable friction between one's strategy and his rival's.

Preferring a winning strategy to a tactical outcome may be a mode of thinking typical to Southeast Asia: military leaders such as Mao Tse-tung in China and Ho Chi Minh in Vietnam won wars without achieving a military decision and without winning major battles. In fact, as opposed to Western military leaders such as Napoleon or George Patton who assumed that given patent tactical superiority one is bound to win any encounter with the enemy force and therefore strategy plays second fiddle, Mao Tse-tung maintained that "the view that strategic victory is determined by tactical successes alone is wrong because it overlooks the fact that victory or defeat in war is first and foremost a question of whether the situation as a whole and its various stages are properly taken into account."[49] This point of view apparently does not stem from the lack of industrial and technological symmetry between East and West, as it is apparent also in wars between different Asian entities, e.g., in the Chinese civil war (on and off, 1927–50).

Nonetheless, most modern Western military doctrines have sought to delimit war to an encounter between the warring militaries in a battlefield for a limited period of time. However, given such delimiting, the advantage that Western armed forces have over third world armed forces is usually clear. Therefore in many cases Asian and Arab military leaders have rejected such delimiting, and have preferred to wage a broader and more holistic confrontation between civilizations, which purposely involves the civilians of both sides, extends beyond the battlefield, and lasts over long or even indefinite periods of time. In contrast to the Western dichotomy, which sees war and diplomacy as two different disciplines conducted with opposite rationales, with different means, at separate times, and even by different branches of the government, the Asian-Arab view blurs the borders between warfare and non-warfare and sees them as a joint synergistic effort. Thus the "the situation as a whole" dampens the techno-tactical advantage of Western militaries and sometimes enables the overwhelming of Western civilian societies, and indirectly, their militaries as well. Moreover, Western democracy is bound by constraints such as considerations of legitimacy, diplomatic freedom of action, and the need to maintain the support of the political, public, and media systems that limit the steps it can take. These systems have a relatively low tolerance level and are far less steady and more vulnerable to enemy manipulations, which makes a "situation as a whole" approach more conducive to the enemies of Western democracies.

## The British Indirect Approach

The trauma of the trench war's pointless destruction on the western front of World War I and a lack of enthusiasm for a frontal attack against the military mass made attacking the enemy's plans a much more palatable choice even among modern Western military thinkers such as, for example, Liddell Hart. In his famous essay on the strategy of indirect approach,[50] Liddell Hart claimed that one of the objectives of strategy is to reduce the opportunity for the enemy to resist it, and therefore the enemy must be surprised in a place, at a time, and in such a manner when it is at its weakest. If the enemy is preparing for any sort of direct confrontation and focuses its strength there, the strategy of indirect approach seeks to avoid that direct confrontation and find another confrontation more suitable to one's relative strength and the enemy's relative weakness. "The real aim [of the strategist] is not so much to seek battle as to seek a strategic situation so advantageous that if it does not of itself produce the decision, its continuation by a battle is sure to achieve this."[51]

If so, this means that victory or decision is not necessarily achieved through the encounter between two sides in the same theater of operations and for the same objective of mutual destruction. As an example, Liddell Hart analyzes the struggle between Britain and Germany in World War I: Britain enjoyed its traditional naval superiority over Germany, but the German ground forces were a tough and bitter opponent. Liddell Hart concludes that on the one hand, the British contribution to the war's success was achieved primarily by the naval blockade imposed by the British on Germany (with minor erosion in only a few naval battles), and on the other hand, the British involvement in the fighting on land made but a meager contribution to Germany's defeat (but exacted a tremendously high price from Britain).[52] The British naval blockade was met by German attempts to sever the naval supply lines between Britain and the United States by using submarines, but the two naval campaigns – the British against Germany, and the German against Britain – were conducted in parallel in different theaters without ever converging.⁵³ (The applicability or lack thereof of Liddell Hart's approach to other nations in different circumstances is discussed in the final chapter of this memorandum.)

A clear example of the British indirect approach may be found in the way Britain undermined Napoleon's campaign in the Middle East and even undermined the strategic level posture of France. In 1798, Napoleon embarked on an expedition of conquests of the Middle East, with his army carried on French navy ships. After Napoleon defeated the Mamelukes in a battle near Cairo, the door was open for control of all of Egypt and even Palestine. Britain refrained from a direct confrontation on land with the French army. Instead, Nelson attacked the French navy ships that had anchored for the night in Abu Qir Bay (in the Nile Delta). The French navy was not only defeated; it was obliterated. Yet more important, Napoleon's army was trapped in the Middle East and lost a basic component of its ability to maneuver in the arena, and France – almost without a navy and with a significant part of its army trapped beyond the sea – lost part of its strategic level posture.

## Scipio Contests Hannibal's Basic Assumptions and Imposes a Different Type of War

The most striking historical example of a victory in war by attacking the enemy's strategic plans, and without the two sides confronting one another in the same theater of operations for the sake of mutual attack of centers of mass, is the Second Punic War between Rome and Carthage, a city-state located on the northern shore of Africa. The Second Punic War began with Hannibal's famous trek through the Alps, which was followed by a series of tactical successes against the Roman legions.⁵⁴ In the second stage of the war, the Roman military leader Fabius Maximus adopted an approach involving avoidance of major battles and denying Hannibal an opportunity for decision, instead choosing to attack his army from the rear and flank in an attempt to draw him into a series of smaller engagements under conditions that lent the Roman side a palpable advantage. (This approach would eventually be known as the "Fabian strategy.") So, for example, the Roman army deployed on forested ridges at the edges of Hannibal's route, and under such field conditions it was hard for Hannibal to make good use of his cavalry, which had given him his primary tactical advantage. Had Hannibal attacked the Romans, the battle would have been fought on their terms, and had he stayed on the open plain, the Romans would not have engaged him in battle. At the same time, the Roman navy was busy severing the supply lines and naval reinforcements from Carthage and New Carthage (Spain) to Hannibal.

However, the most dramatic stage of the Second Punic War began

## ONE WAR WITH TWO CONVERGENT CAMPAIGNS

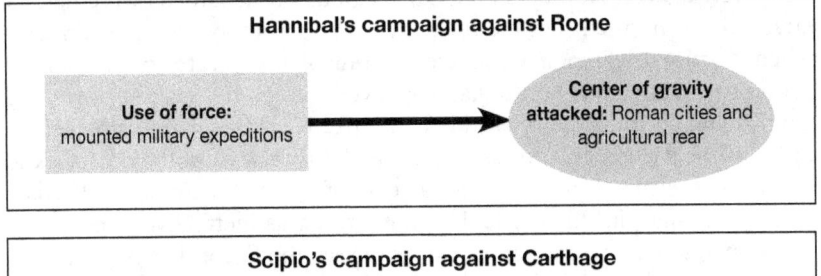

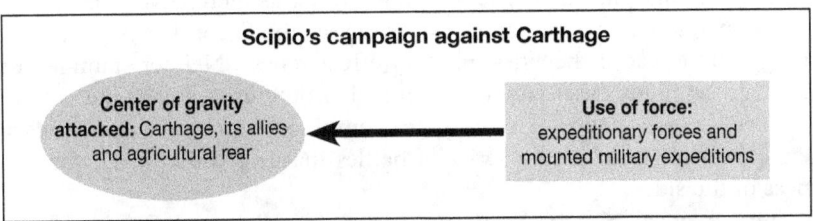

Figure 5.1 The Second Punic War: Two parallel non-convergent campaigns

when the Roman military leader Scipio Africanus took the reins in hand. Scipio decided to avoid confrontation with Hannibal on Italian soil altogether, and instead set sail first for New Carthage (Spain), where he attacked Hannibal's main base, and then continued to North Africa, threatened Carthage itself, destroyed its agricultural rear, and attacked its allies, thus in effect attaining a decision in the war (figure 5.1).[55]

Scipio's greatness lay in his success in attacking Hannibal's basic assumptions. Formulating a war paradigm that would benefit him irrespective of the choice Rome made, Hannibal shaped a war that ostensibly gave Rome two alternatives: take positions against him in the tactical battlefield and suffer defeat after defeat, or give him the freedom of action to sack Rome's agricultural rear. These alternatives rested on the "self-evident" basic assumption according to which there was only a single theater of operations – Italy – in the Second Punic War. Scipio contested this basic assumption, opened additional theaters of operations, and in effect left Hannibal to face two alternatives of his own, both of which benefited Scipio: abandon Italy and take positions for a battle for Carthage, or give him freedom of action against the capital city. Contesting the enemy's "self-evident" basic assumptions, refusing to operate within the range of the alternatives the enemy presented, and imposing a set of different alternatives – each of which was good for Rome – underlay Scipio's winning strategy. Indeed, the highest level of military leadership lies in making the enemy wake up one morning to a war whose characteristics and nature the enemy had not contemplated or planned.

Scipio won the Second Punic War, though one can claim that the victory did not result from a blow to Hannibal's capability to operate

## PARALLEL WAR

effectively against Rome – Hannibal's legions on Italian soil were neither attacked nor harmed. On the other hand, it is also possible to claim that when Scipio threatened Carthage, Hannibal lost strategic freedom of action to act effectively on Italian soil, even though he was not prevented from doing so from the operational perspective. Hannibal's armed force did not lose its "generic" operational capability, but from the outset it did not have the necessary operational capability to foil Scipio's alternate strategic design; in this particular context, it was ineffective almost as if decision against it was attained. Nonetheless, the Second Punic War still hinged on a test of the sides' military effectiveness. Neither stamina nor resources nor any other factor determined its outcome, only Rome's effectiveness in conducting a long-range, combined naval and land military expedition that included a series of battles that tested the field effectiveness of the sides.

One may call Scipio's military doctrine "parallel war." Scipio refused to appear at the campaign that Hannibal chose to conduct, whose rationale and nature were suited to the strengths of Hannibal's forces and the weaknesses of the Romans. He chose to embark on a parallel campaign whose rationale was to threaten Carthage's strategic centers of gravity directly – the city of Carthage itself, the agricultural rear feeding the city, and its allies. This threat enabled Scipio to impose an end to the fighting on his terms and from a position of military-strategic dominance – in other words, to achieve a strategic surrender.

In parallel war, each side embarks on a separate campaign devised to favor its relative strengths and the opponent's relative weaknesses, and tries to attack different enemy centers of gravity. Each side seeks to realize its campaign plan with the intensity and effectiveness that will enable it to be the first to achieve two requisite goals:

1 The campaign conducted must strike or threaten the enemy's centers of gravity, which will enable that side to impose an end to the war on its own terms.
2 This also includes denying the enemy its strategic freedom of action to continue fighting the parallel campaign it is conducting (even though it may still have operational level freedom to do so).

These two achievements together create the strategic superiority and dominance needed for victory in parallel war.

## Fundamental Concepts Reexamined

A survey of the cases, terms, and analytical tools examined in the earlier chapters (including primary test cases, such as the Yom Kippur War on the southern front), the expanding of the doctrinal background surveyed at the beginning of this chapter, and the parallel war as illustrated by the Second Punic War oblige us to refine the definitions used in wars testing the military effectiveness of the sides, beyond the definitions already existing in the various doctrines. The most important distinctions are between operational centers of gravity and operational decision, involving a blow to the enemy military's operational capability; and the strategic centers of gravity and strategic surrender, involving denying the enemy strategic freedom of action to continue fighting at all, or at least to continue fighting a war conforming to the paradigm suited to the enemy. In this context, this memorandum suggests some new definitions, as follows:

*A military-operational center of gravity* means a place, force, function, capability, basic assumption, plan, or any other component of the enemy force located in or connected to the operational space of the war, and is identified by the fact that its attack causes the rest of the enemy system's components in the operational space – themselves not under direct attack – to stop operating effectively in realizing their designated function in terms of the relevant level, context, and circumstances.

*Military-operational level decision* is achieved by striking the enemy's military-operational level center of gravity, and is identified by the fact that the enemy force constellation concentrated in the theater of operations has lost its capability of operating effectively, in terms of its ability to realize its campaign theme and its strategic objective, considering the context and the circumstances, and the fact that the chances of the enemy force recovering from this strike and adapting to the new situation in the course of the same war are very low.

*Strategic center of gravity* means a place, force, function, capability, basic assumption, plan, or any other component of the enemy system as a whole, not necessarily located in or connected to the operational space of the war, which if struck would bring about (or if threatened is likely to bring about), directly and in a time frame relevant to the circumstances, a denial of the enemy's strategic freedom of action to continue fighting, or at least a denial of the freedom to continue fighting according to the enemy's preferred war paradigm.

*Strategic surrender* is achieved by striking or threatening the enemy's strategic center of gravity, and is identified by the fact that the enemy seeks an immediate exit from the war because of distress (i.e., not in order to protect its gains). Ideally the strategic surrender is accompanied by a military operational level decision, the destruction of the mass of the enemy force, and the demonstration of superiority at every level of the war, but it may occur for lack of any other option even if the enemy force maintains its capability to continue operating effectively at the field levels.

Table 5.1 charts a matrix linking the nature of the war, its issues, examples of the main force application thread, and examples of centers of gravity under attack.

**Table 5.1** The Nature of the War, Force Application Thread, and Centers of Gravity

| The nature of the war | The issue the war hinges on | Examples of main force application thread | Examples of centers of gravity under attack | Comments |
|---|---|---|---|---|
| Decision | Military effectiveness of the sides | Direct or indirect blow to the enemy's military capability | Central mass or weakness in the military formation | Attacking the center of gravity is supposed to affect the military course of the war within hours or days |
| | | Denial of strategic freedom of action to fight | Strategic; enemy plans or paradigm | |
| Other than decision (lack of decision within a short period of time) | National stamina | In a dictatorship: a blow to the leadership's will to fight | Raising the cost of the war (in the currency dearest to the leadership) | The asset under attack is not in and of itself a center of gravity. An ongoing attack of assets is supposed to affect the center of gravity indirectly (within weeks or months) |
| | | In a democracy: a blow to the civilians' willingness to support the war | The trust and cohesion between the government, civilians, and military | |
| | National war supporting resources | A direct strike against the enemy's resources | The enemy's war supporting industries | The asset under attack is itself a center of gravity, but its effect on the course |

# 6

# The Second Lebanon War and Operation Cast Lead

## Parallel Wars against a Non-State Opponent

This chapter analyzes the Second Lebanon War using the analytical tools and the terminology clarified thus far, and claims the following:

1. Hizbollah's paradigm of war consisted of attacking the stamina of the Israeli political–civilian system (a strategic center of gravity) using two alternate methods: directly, via rocket fire, and indirectly, using guerillas against the IDF in southern Lebanon. The Hizbollah plan seemingly enabled Israel to choose only which soft underbelly to expose.
2. Israel operated within the framework of Hizbollah's paradigm, in particular by assuming positions for a parallel war in which fire is exchanged over an extended time period in a way that tested the stamina of both sides (a test that Israel is not really interested in).
3. Israel operated according to the American mode of avoiding attrition by avoiding a confrontation on the ground in the tactical battlefield, without understanding that given the particular circumstances and Israel's geo-strategy, leaving the enemy with tactical capabilities enables the enemy to engage in reciprocal strategic attacks. This creates a reciprocal test of staying power, and within the Israeli frame of reference that amounts to attrition.
4. In order to realize its plans and be effective at the higher levels of war, Hizbollah was required only to show residual defiance capability. Even inflicting a severe tactical blow to Hizbollah would not have denied it the capability of operating effectively at the higher levels of the war. Given this reality, Israel did not fully work out what were the requisite elements for achieving decision over Hizbollah, namely, how to prevent it from acting effectively to realize its plans at the higher levels of war.
5. Hizbollah made its physical centers of gravity more amorphous. Israel did not fully determine which of Hizbollah's centers of gravity it

should aim to attack, and certainly did not specify those centers of gravity that if attacked would advance the rationale of a war of decision (as opposed to a war of attrition/stamina).
6. Hizbollah's ability to continue attacking Israel's stamina over time using residual defiance capability, along with its concealment of its physical centers of gravity, are the basis for the phenomenon of "winning by not losing," allowing it to outlast Israel. This phenomenon seeks to avoid a quick test of the sides' military effectiveness and extend the war until other issues, such as stamina and staying power, determine the outcome.
7. Israel saw the war as no more than a list of targets to attack with standoff fire. The failure lies not only in the lack of effectiveness of standoff fire in the particular circumstances of that war, but also in the very perception of the war as a list of targets to attack. Israel did not define which issues it sought to test in the war, and which force application thread would lead the war towards those issues.

## Hizbollah Attacks the "Spider Web"

The most prominent recent example of an asymmetrical, multileveled, multi-issue, parallel war challenging the basic assumptions of classical doctrine is the 2006 Second Lebanon War. Israel's strategic center of gravity that Hizbollah sought to attack was what Hizbollah secretary general Hassan Nasrallah has called the Israeli "spider web," in other words, the perception that an affluent, Western democratic society is unable or unwilling to sacrifice and pay the price of an ongoing confrontation, and therefore sooner or later it withdraws and concedes (in fact, this center of gravity belongs to the grand strategy level, but it is commonly referred to as a strategic center of gravity). The concept of the spider web is somewhat reminiscent of Clausewitz's "wondrous trinity," which maintains that in order to conduct a war, it is necessary to harness together the government, the military, and the people in a joint effort. To use a different formulation (that goes beyond Clausewitz's definitions, which are more suited to the socio-political map of the early nineteenth century), one may define the Israeli center of gravity that Hizbollah chose to attack as the trust and cohesion between the government, the military, and the civilians.

Hizbollah's capacity for operating effectively against Israel's strategic center of gravity was made possible through the confluence of particular circumstances, unparalleled in the annals of wars between states and non-state opponents. First, regular armed forces would usually meet irregular or non-state forces across the sea, so that the operational capability of the non-state force against the strategic center of gravity of the state opponent was limited to indirect attacks. The Vietcong could not attack the American home front directly, and the Malayan insurgent organization MCP could not attack Britain directly. It is true that Algerian groups carried out some attacks against French civilians, but their ability to operate within France itself was limited and only rarely expressed. While Russia has a common border with Chechnya, which was breached on rare instances to attack the Russian home front, the geographical distance from the main Russian population centers to the Chechen border has created a reality similar to a war across the sea. Therefore, the more common mode of operation of the various guerilla organizations is to strike at the state opponent's military expeditionary forces, and thus indirectly pressure its strategic–civilian–political center of gravity.

However, the geographical proximity of southern Lebanon to major Israeli population centers gave Hizbollah (and, prior to that, the PLO) the ability to turn simple cheap short range rockets into a strategic weapon. Israel's unique geo-strategic position – i.e., the lack of depth between the front and the rear – means that tactical weapons are capable of directly inflicting strategic blows. In a manner that is unusual for wars between states and guerilla organizations, Hizbollah thus had the capability of attacking an Israeli strategic center of gravity, the home front, (1) directly, (2) intensively, (3) continuously, and (4) whenever it wanted.

A second unusual feature was that Hizbollah combined military capabilities comparable to those of a state (such as the number of stockpiled rockets and the capacity to strike the strategic rear of the state opponent, as well as the quality of its anti-tank weapons) with the ability to disappear in the face of intelligence gathering and maneuvers and to survive fire, which are typical of non-state guerilla organizations (blending into the local civilian population, having a low signature, secrecy, and compartmentalization, and showing minimal sensitivity to harm inflicted on the host country (during the course of the war).[56] Along with the ability to disappear, Hizbollah was also capable of hardening its positions against fire using higher signature methods, especially in the network of its "nature reserves." This feature enables us to view Hizbollah as a quasi-regular fighting force, though still as a non-state entity. In addition, Hizbollah was an authentic representative of the Shiites of southern Lebanon, and therefore could not be easily uprooted. A partial list of the salient features pertinent to both Lebanon Wars appears in table 6.1.

Table 6.1 Distinctive Attributes Separating the First and Second Lebanon Wars

| PLO, 1982 | Hizbollah, 2006 |
|---|---|
| Foreign in Lebanon (except in the Palestinian refugee camps) | Authentic representative of the local Shiite community |
| Operating in large, semi-military formations | Guerilla-like operation |

| | |
|---|---|
| In response to IDF maneuver, organized withdrawal northwards while maintaining formations and a relatively high signature | In response to IDF maneuver, disappearing and blending in; continuing guerilla warfare in southern Lebanon over time |
| Taking positions for a series of relatively large battles | Usually, an avoidance of prime major battles |
| The maneuver afforded an opportunity for decision | The maneuver does not afford an opportunity for decision |
| End state: final and absolute uprooting from Lebanon | No possibility of final and irreversible uprooting; after withdrawing, the situation is liable to revert to its previous state |

Against the background of these particular circumstances, Hizbollah developed a two-pronged approach – direct and indirect – to attack Israel's strategic center of gravity (figure 6.1). The one, enabling direct attack on Israel's center of gravity, was the rocket force capable of generating intensive and prolonged fire on the civilian home front. Beyond the (limited) physical effect of the rockets, their operation destabilized the Israeli center of gravity in several respects. First, the Israeli government at times found it difficult to provide services (food, medical care, and so on) in the areas under fire, which impacted on the cohesion of the government-civilian axis. Second, the civilians did not receive the level of protection they had expected from the government, and that too impacted on the cohesion of the government-public axis. Third, the IDF did not deliver the military results that the government expected (suppression of Hizbollah rocket fire), and this impacted on the cohesion of the government–military axis. Thus, the center of gravity was not the direct damage inflicted by the rockets themselves, rather the ramifications of their firepower on the structure of the Israeli political–civilian system.

Hizbollah's rocket force was constructed in a way that combined state capabilities with the guerilla organization's low signature and ability to disappear, and so it could not be neutralized using standoff fire,[57] but only through conquest and systematic clearing of southern Lebanon (therefore, ideas suggested in Israel such as cutting southern Lebanon off from the north or conquering only some high grounds were irrelevant). In case Israel would choose to occupy southern Lebanon, Hizbollah established a fighting guerilla force along with its rocket force whose purpose was to exact a high cost in casualties from Israel for the duration of the occupation, thereby indirectly attacking Israel's strategic–civilian–political center of gravity (i.e., operating on the basis of the more typical guerilla model). In other words, even though Hizbollah's ground force was designated for use in the battlefield, its purpose was not to halt the IDF on the operational level (and given its understanding of its relative strengths and weaknesses, not even to deploy for big battles), rather to create a situation that was intolerable over time yet lacked a military exit, thus indirectly undermining the will of the Israeli public and the

## PARALLEL WARS AGAINST A NON-STATE OPPONENT

political system to fight, much akin to the asymmetry between the United States and the Vietcong discussed above.

It is hard to claim that Hizbollah was indifferent to the possibility of Israel occupying southern Lebanon. Nonetheless, it was built to generate an operational level and strategic level advantage from either of the two possible situations. Moreover, since Hizbollah enjoyed the capacity, typical of a guerilla organization, to blend in and disappear with the benefit of local popular support, the concern arose that an end to the occupation of southern Lebanon would result in the situation being restored to its former state. Therefore, shaping a successful exit strategy for an occupation of southern Lebanon became a complex issue, and the occupation might have been liable to continue for an extended period with no end date in sight. Under these circumstances, in which an occupation would have been seen as lacking both an end date and a useful purpose, Hizbollah's capability of striking at Israel's strategic center of gravity by harming Israeli soldiers in southern Lebanon would only grow.

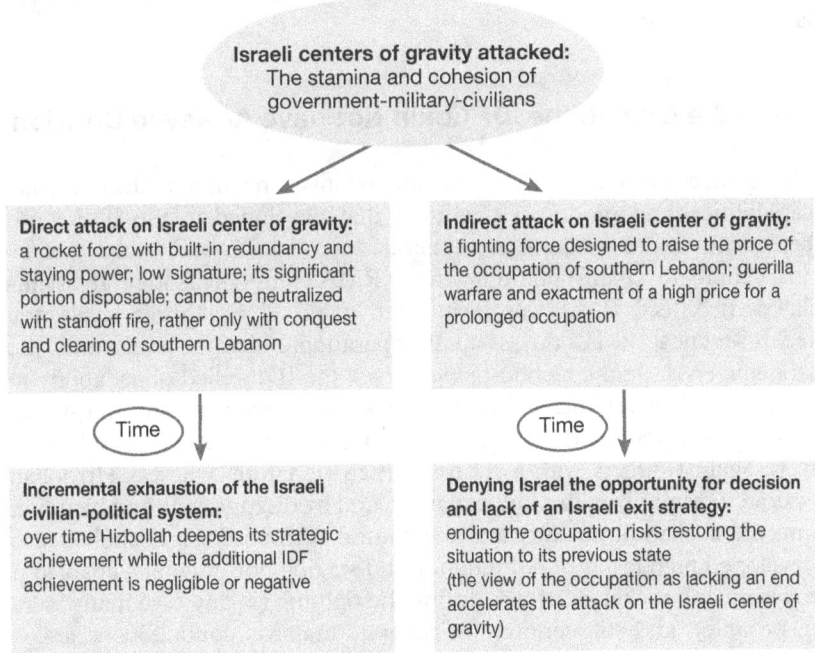

**Figure 6.1** Hizbollah's War Paradigm

Thus, Hizbollah presented Israel with two bad choices: expose the Israeli home front to direct, massive, and prolonged strategic fire, or occupy southern Lebanon and be worn down by a prolonged, unpopular occupation lacking a definite ending. Moreover, in order to achieve these operational and strategic level outcomes, Hizbollah did not need to maintain its military capabilities fully. For example, even if the IDF had succeeded in inflicting a significant blow to Hizbollah's operational capability and reduced the number of rocket launches by 60 percent – from 250 a day to 100 a day – Hizbollah would still have succeeded in eroding the cohesion between the government, public, and IDF over time, because the disruption to the civilian life in Israel and the sense that the home front was defenseless would have been achieved to a similar degree even with the lower number of launches. As long as Hizbollah retained its residual defiance capability vis-à-vis the Israeli home front, it would have gained its desired outcome (sirens, descent into bomb shelters, evacuations, disruptions of civilian life). The index for measuring Hizbollah's operational capability is its ability to obtain the desired outcome at the higher levels of war, and that was only slightly influenced by the outcome at the tactical level, i.e., by the question of the number of launches per day and how many launchers it lost.

## From the Outset, the IDF Could Not have Achieved Decision

Hizbollah constructed itself in a manner relatively resistant to the rationale, capabilities, and modes of operation that developed in the IDF, both because of its very nature as a guerilla organization and because of its adaptability to a confrontation with the RMA components the IDF assimilated. In effect, RMA was assimilated in the IDF at three levels: the techno-tactical level, the systemic-operational design level, and the strategic level. At the techno-tactical level, the IDF relied on its ability to gather intelligence and attack targets in a standoff manner using long-range means. However, the IDF's capabilities of intelligence gathering and attack were against targets with a relatively high signature, whereas Hizbollah operated primarily with a low signature, and by disappearing and blending into the civilian population and the natural surroundings. Hizbollah also developed high levels of redundancy and staying power, which enabled it to lose many rocket launchers during the fighting (in any case many were disposable), and still continue to generate massive, continuous strategic fire against the Israeli home front. In addition, Hizbollah also hardened its assets to be able to withstand standoff fire, primarily by moving assets into underground systems.

At the systemic-operational design level, the IDF sought to analyze its enemy as a system of systems, to identify vulnerability nodes in the opposing system, and to attack them in a way that would generate effects suppressing the functional effectiveness of the rival system. However, Hizbollah tried – and to a great extent succeeded – to construct itself as an "organization not functioning as a system" (to the extent possible, of course, as it is indeed a hierarchical organization). It adopted a flat, decentralized structure consisting of a network of local cells operating almost autonomously on the basis of instructions given in advance and very few simple, short instructions given in real time to local and even junior commanders who enjoyed extensive discretion. Hizbollah pre-positioned a large part of its fighting forces, means of warfare, and supplies in such a way that it hardly had to transport personnel or operate a logistical system in the course of the fighting. The internal interaction required by the Hizbollah organs in the course of the fighting was minimal.

At the strategic level, the IDF sought to attack its opponent's strategic centers of gravity by using long-range high output firepower to illustrate to the enemy the cost of the war and to bring about the loss of the will to fight and cognitive dissolution within the enemy's leadership. These centers of gravity were understood by the IDF primarily in the physical sense, i.e., valuable strategic assets. However, barring some notable exceptions, such as the Dahiya quarter in Beirut, which served as a command and control center, a residential neighborhood for the families of the organization's leaders, and a symbol for Hizbollah's strength and (defiant) autonomy within the Lebanese system, Hizbollah had very few valuable physical strategic assets. The IDF did attack Dahiya after issuing repeated warnings to civilian residents to evacuate their homes. Attacking Dahiya, therefore, was no surprise and did not deal a significant blow to Hizbollah's leaders or to its command and control capability. However, attacking the quarter definitely had symbolic meaning, and more importantly, the fact that Israel attacked semi-civilian residences in the heart of Beirut – not by mistake or on a one time basis but systematically, over and over again – represented something of a break in Hizbollah's war paradigm. The Dahiya quarter was undoubtedly an important asset of intrinsic and symbolic value to Hizbollah, but was it a center of gravity? Attacking Dahiya did not impair Hizbollah's fighting capability, and it seems that it had no effect on the organization's strategic freedom of action to continue fighting. These facts raise the following question: in the context of a war of decision (rather than a war of attrition or stamina), was it correct to categorize the Dahiya quarter as an essential strategic center of gravity (at least regarding its occupation after the warnings about the impending attacks had been issued to the residents – and thus also to the Hizbollah leadership)?

The senior Hizbollah leadership certainly represented a physical strategic center of gravity, but it is incorrect to base war plans on the assumption that it will be possible to locate and attack individuals (just as the Americans failed to locate Saddam Hussein at the high intensity stage of the 2003 Iraq War or Osama Bin Laden since 2001). Hizbollah also had abstract strategic centers of gravity, but Israel did not operate in a coherent systematic manner on the operational level in order to attack them, and did not connect such operational designs with a strategic–political end state. Thus, Hizbollah constructed itself within the blind spot of IDF capabilities and ways of thinking (table 6.2).

Table 6.2 Hizbollah Deconstructs the IDF's Capabilities and Plans

| The IDF capability/plan | Basic assumptions for operating effectively | Hizbollah's response |
|---|---|---|
| **At the techno-tactical level:** Operating precise, rapid standoff fire at enemy targets. | Large bank of high quality targets, capability of generating additional high quality targets swiftly. | Operating with a low signature, disappearing and blending in among the civilians and into the natural surrroundings, usually not congregating to form targets, high redundancy rates, hardened and tunneled positions. |
| **At the operational level:** Suppressing the enemy's effectiveness as a system of systems. | EBO requires an enemy (1) organized as a system; (2) having vulnerability nodes whose attack will suppress the system's effectiveness; (3) good intelligence contact with vulnerability nodes. | "An organizaton that does not function as a system": a flat, decentralized organization constructed as a network of autonomous cells, forces and assets deployed ahead of time, and a command and control system that can disappear and has built-in redundancy; compartmentalization and secrecy lead to low levels of familiarity. |
| **At the strategy level:** Long-range fire attacks against the enemy's strategic centers of gravity. | The existence of strategic centers of gravity whose attack will impair the enemy's fighting capability or strategic freedom of action to continue fighting. These centers of gravity were understood primarily as assets with physical characteristics. | The lack of clear physical strategic center of gravity, whose attack would bring about the collapse of Hizbollah (a question mark with regard to the Dahiya and the senior leadership). |
| **Rapid decision capability:** Intensive strike capability based on aerial firepower. | The existence of physical and accessible centers of gravity against which force is applied. | Thinning out and obscuring physical centers of gravity, focusing on developing own stamina and challenging the stamina of the Israeli home front, in order to shape a war that revolves around staying power. |

The Second Lebanon War also illustrates the incongruity of some of the notions that developed within the IDF in recent years regarding the nature of war suited to Israel. The ideas of standoff firepower-based operations, decentralized warfare, and the dynamic molecule (once carried out as the main mode of battle[58]) are particularly apt for a side seeking to disappear from the theater of operations, extend the duration of the war, wear down the opponent with firepower, and put the sides' stamina to the test. These notions may well be suited to guerilla organizations or weak countries seeking to avoid a battle of decision and exhaust the enemy with continuing defiance, but they do not suit a nation seeking to test the military effectiveness of the sides in a short period of time. A military seeking to prove superiority on the battlefield cannot disappear from it. Moreover, as Israel learned during the last war and as the United States learned in Vietnam, the sensitivity of guerilla organizations and third world dictatorships to effects directed at the political–civilian system is relatively low. In contrast, open democracies and affluent, industrialized nations such as Israel and the United States are much more vulnerable to such effects. Indeed, the asymmetry is also noticeable in Hizbollah's success at generating strategic effects against the civilian population, the political system, and the Israeli economy, while because of the nature of the situation, Israel could not enjoy similar recompense.

## Israel Plays the Part Scripted by Hizbollah

Similarly, during an interim stage of the Second Lebanon War, the IDF started carrying out special forces and infantry raids against Hizbollah targets in southern Lebanon. However, the raid as a primary form of battle (rather than as a measure in aid of a different effort) also falls under the rationale of an exchange of blows, which in the end tests the sides' stamina. Raids generally do not contribute to shaping a stable, reality-changing military end state; rather, they are consistent with a ping-pong match of mutual bloodshed and harassment, i.e., a struggle over stamina and attrition. Raids cannot be a main force application thread in a war of decision, testing the sides' military effectiveness in a quick burst; this requires the application of maximum force in order to realize the sought-after end state (defined beforehand) in the minimum of time.

In truth, Israel did not grasp the nature of the war it fought with Hizbollah, and did not succeed in attacking the organization's plans or imposing on it a war of a different nature and rationale. Therefore, Hizbollah's structure and the nature of the war it crafted resulted in the loss of effectiveness of the IDF's paradigm to a great extent even before the first shot was fired. Nor did the IDF have a strategy or an approach before the war broke out that would enable it to operate effectively against

Hizbollah. Israel also did not define for itself in a coherent fashion what decision against Hizbollah meant, and what components of decision it sought to achieve at the military end state. The continuing erosion of the cohesion of the Israeli triangle of government–military–civilians, and as a result the erosion of Israel's stamina using only Hizbollah's residual defiance, while Hizbollah's physical centers of gravity and mass were concealed and disappeared, are the basis of the phenomenon of "winning by not losing." In order to achieve a decision against Hizbollah directly, it was necessary to deny it even its residual defiance capability, and that is a lofty and at best difficult goal indeed. Furthermore, the option of winning by not losing was never open to Israel. This option is available only to a side seeking to avoid a decisive test of the sides' military effectiveness, and is interested in extending the fighting until other factors, such as stamina, staying power, and the international community determine the outcome of the war. Israel has the potential for superiority of military effectiveness, and an inherent inferiority in almost every other test of war, and therefore cannot win by not losing.

In fact, Hizbollah invited Israel to wage a war based on mutual exchanges of fire, testing the stamina of the sides over time, and Israel – surprisingly – accepted the invitation and put its stamina to the test, a test it was not really interested in. The reluctance to be mired in a conflict of attrition on the tactical battlefield of southern Lebanon produced attrition through exchanges of fire, with Israel's strategic centers of gravity open and accessible and Hizbollah's strategic centers of gravity hard to identify and access, and with Israel operating under more pressing limitations in terms

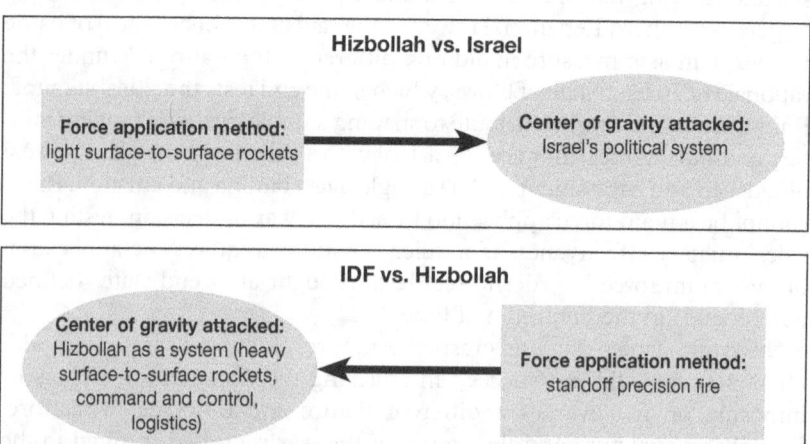

**Figure 6.2** The Second Lebanon War: Two parallel, non-converging campaigns

of political, diplomatic, and economic staying power. In this reality, where each side attacked the selected centers of gravity of its opponent with fire, there were in fact two parallel firepower-based campaigns: Hizbollah's campaign against Israel, and the IDF's campaign against Hizbollah. Hizbollah attacked the Israeli civilian home front, whereas Israel tried to attack Hizbollah's vulnerability nodes as a system as well as identify and attack strategic centers of gravity that would affect it (fig. 6.2). These two campaigns hardly ever converged in the same theater of operations (with the exception of limited – and belated – Israeli maneuvers). Moreover, on the tactical and operational levels, the two campaigns hardly ever disrupted one another, and barely affected one another in any significant way.[59]

The moment a parallel war of this nature erupted, the IDF was forced to conduct two campaigns at the same time based on different systemic-operational rationales and modes of action. On the one hand, it was required to achieve an absolute success in its campaign against Hizbollah in order to realize the objectives of the war prescribed by the Israeli government. On the other hand, it was required to achieve a relative success of disrupting the opposite campaign – Hizbollah against Israel – to such an extent as to not allow Hizbollah's success to eclipse Israel's success in its campaign against Hizbollah. The IDF failed twice: first, it did not win the Israeli campaign against Hizbollah, i.e., it did not succeed in operating military force in a way that would realize the political objectives of the war, and second, it failed to prevent Hizbollah from operating effectively in its campaign against Israel to any degree worth mentioning.

Many of those who criticized Israel's management of the Second Lebanon War, including this author, claim that the root of the failure lies to some extent in the deterioration of the IDF's maneuvering capability in favor of a view that war is nothing more than a process of producing pinpoint accurate targets and their attack by precision standoff firepower. However, even if the IDF had embarked on a limited maneuver in southern Lebanon (given the circumstances that prevailed at the time), the most success it could have hoped for was disruption of Hizbollah's firepower-based campaign against Israel, and trading it for an exhausting anti-guerilla campaign. A limited maneuver in southern Lebanon, no matter how successful, would not have been enough either to stymie Hizbollah's attainment of its war paradigm or to attain victory in the parallel campaign conducted by Israel against Hizbollah: that is, realizing the objectives prescribed by the Israeli government, specifically the permanent removal of the Hizbollah threat against Israel and its dismantlement as an armed player on the Lebanese stage. Hizbollah had no center of gravity vulnerable to a maneuver that would bring about the collapse of the other components of the organization, and certainly not a permanent collapse without the possibility of recovery. Unlike the PLO in 1982 (which was pushed northwards and permanently expelled from Lebanon), Hizbollah did not operate in a mode that would have led to its expulsion as an organization from southern Lebanon, but only to its disappearing and blending in with the local Shiite population and its transition to guerilla warfare against the IDF's supply lines and occupying force. Unlike the PLO of 1982, which was an alien entity in Lebanon and therefore liable to be entirely uprooted, Hizbollah is a grassroots expression of the Lebanese Shiia and therefore cannot be uprooted. Occupying southern Lebanon might therefore have become a reality without an exit strategy, as after a withdrawal, the situation might easily have reverted to its earlier state.

Therefore, from the very outset Israel was incapable of winning the Second Lebanon War within the nature and rationale of the war as dictated by Hizbollah. Only a strategic attack (not necessarily physical) on Hizbollah's plans and basic assumptions, or an attack on its abstract centers of gravity or centers of gravity falling outside the boundaries of the campaign conducted in actuality, and imposing a war with a different nature and rationale, might perhaps have produced decision and a victory.

## Could it have Evolved Differently?

Clausewitzian analytical tools, RMA analytical tools, and the conceptual approaches that took root in the IDF in recent years proved insufficient for the complexity of the Second Lebanon War. Consequently, what analytical tools and what thought processes were required in order to plan and wage the war against Hizbollah? This chapter has established that the starting point had to be identifying Hizbollah's abstract strategic centers of gravity and planning how to take advantage of an attack against them to achieve the desired political–strategic end state – a victory. One possible move was to refuse to accept the paradigm whereby Israel faced a non-state opponent with few physical centers of gravity, and instead to expand the war's boundaries to involve state players as well – Syria, Iran, or Lebanon.

As for Syria and Iran, Israel chose to delimit the war and not include those two countries. One may criticize that decision, though Hizbollah plays merely a supporting role in the system where Iran and Syria are much more dominant, heavy-hitting players. A war involving Iran or Syria must first of all take them into consideration and be based on a strategy that is coherent in terms of dealing specifically with them and not as an indirect means for dealing with Hizbollah. Including Iran or Syria as an afterthought in a campaign having Hizbollah at its center is a case of the tail wagging the dog. It may well be that a military confrontation with these countries has its place, but not at the timing and within the context of Hizbollah's choosing.

Nevertheless, another state player could be included – Lebanon. The abstract strategic centers of gravity of the Lebanese system highlight Hizbollah's place within that system. After the withdrawal of the foreign forces – Syria and Israel – from Lebanon, the legitimacy of Hizbollah as an armed player on the Lebanese stage became the organization's prime concern and defined its relations with the other components of the Lebanese system. If in the context of "resistance" to Israel the armed Hizbollah was able to justify its existence as an asset to the Lebanese state, then after the withdrawal of the foreign forces it might be seen as a burden. This tension is heightened in particular under circumstances where Hizbollah is seen as applying its force in favor of foreign interests (Iranian or Syrian) and not on behalf of Lebanese national interests, and especially when the Lebanese state is paying a steep price in the services of those foreign interests. Thus the critical question is whether Hizbollah as an armed entity represents a valuable component of the Lebanese state order, or conversely, does it represent a threat against the state order, which is unstable to begin with? A different Lebanese internal center of gravity feeding Hizbollah is the oppressed and adversarial status of the Shiite community in Lebanon. Here, the question tested is whether Hizbollah is the local Shiite knight in shining armor, or does its activity exact an intolerable price?

Israel's military activity in the Second Lebanon War highlighted the overt and hidden tensions between Hizbollah and the other factions in Lebanon, and the tensions inherent in that Hizbollah, as an Iranian proxy, exacts a cost of the Lebanese state. The military activity also revealed fissures between Hizbollah and the moderate Sunni Arab nations. However, these questions were not raised in an attempt to realize a coherent systemic-operational theme or to connect them with any end state. Could this have been done differently (at least in the reality of 2006)?

The first point requiring clarification is whether and how, if at all, it was possible to apply military force along with non-military means and build a coherent and practical situation demonstrating that Hizbollah is a burden on Lebanon, a threat to its state order, and perhaps even a burden on the local Shiia. The challenge lay in the capability (or lack thereof) of translating these abstractions into a concrete and achievable operational plan.

The second point is whether it was possible to connect such an application of force directly to a political–strategic end state. Both the Lebanese government and the Lebanese armed forces are notable for their weakness, and therefore it is not clear whether it would have been possible to push them into a confrontation with Hizbollah, and if so, what its outcome would have been. (This is with reference to the reality of 2006; in the current reality, the probability that the Lebanese government or military would take a stand against Hizbollah is very low and Hizbollah has in any case become a dominant force with veto rights in the Lebanese government, pushing the Lebanese armed forces out of bounds of Hizbollah interests.) Significantly,

Israel should have understood that sooner or later war would break out against Hizbollah, and in this particular case, Israel's national interests would be seen as legitimate and congruent with the interests of the Lebanese government, the United States, and France, and even with the interests of the moderate Sunni Arab nations. (The world consensus was obvious, for example, in the G-8 statement of July 16, 2006.) Therefore, it was both possible and correct to begin formulating a political end state even before the outbreak of the Second Lebanon War, and Israel should have contributed its part to this end state by responding positively to some of the Lebanese government's demands. Later, when the war broke out, applying military force should have created the context in which the strategic–political end state could be imposed. However unlike 1973, the connection between the application of military force and the end state had to be much stronger. While it may have been possible to attain different components of the political end state even without a direct causal connection to the military end state (for example, with regard to the conduct and motivation of the Lebanese government in its 2006 configuration), the same cannot be said regarding the disarming of Hizbollah's military capabilities.

Therefore, the third point is that it cannot be expected that any element, Lebanese or foreign, would be willing to or capable of dismantling Hizbollah's military system in the south, and only a conquest of the region from the border to the Awali River (and perhaps even north of it) by the IDF and clearing it thoroughly would have enabled the political end state (including transferring the cleared areas into the hands of the Lebanese military). Thus, the application of military force should have served two axes: a strategic axis that would demonstrate that Hizbollah, as an armed organization, represents a burden on Lebanon, thereby creating the context that would have enabled the Lebanese government to play its part (again, in the reality of 2006, not of 2009), and an operational axis, meant to impact on Hizbollah's capability of acting against Israel and also to physically prepare the political end state (which was to have been determined ahead of time). The strategic axis is meant to realize Israel's war objectives (victory – asserting the political responsibility of the Lebanese government over its entire territory and over Hizbollah), and the operational axis is meant to foil Hizbollah's opposite campaign (decision), and also prepare the end state from the physical perspective. In such a multilevel context, it is possible that Israeli maneuvering would have had positive value, and it may even have been able to suggest a relevant exit strategy.

A fourth point is that the systemic-operational theme of occupying southern Lebanon and clearing it should have presented Hizbollah with challenges exceeding its capabilities. One example: as a small organization that at the time had only hundreds of first-rate fighters, it would have found it difficult to withstand a massive, multi-focal effort on a wide and deep front that was sustained continuously over many days.

## The Real Lessons from Lebanon

This monograph does not present a definitive opinion on the practicality and applicability of these four points to the reality of 2006 or 2009, and it is not self-evident that it would have been possible to turn such ideas into an operational plan and a stable, long lasting end state. However without a doubt, before launching the war the government of Israel should have been convinced that adequate positive answers existed to these and other similar questions. One does not go to war without first clarifying the picture in its entirety, from the end to the beginning. One of the primary lessons summarizing the failures of the Second Lebanon War is the emphasis that going to war always requires an organized process in which first of all, the desired political–strategic end state is defined, and its connection to the military end state is defined. The process of characterizing a war obligates a definition of the issues to be tested in the war, such as the sides' military effectiveness or other selected issue (resources, stamina, ability to enlist the international community, and so on). The selection of issues to be tested by the war is always context-dependent, and entails an assessment of one's relative advantage vis-à-vis the specific enemy and under the prevailing circumstances.

At that point, the main force application thread must be defined so that it will channel the war towards the selected issues. In wars seeking to test the sides' military effectiveness, this thread is likely to be, for example, a direct or indirect attack on the enemy's fielded forces (Israel in the Six Day War), or denying the enemy its freedom of action to continue fighting (Scipio's threat against Carthage), or deconstructing the war plans of the enemy forces (Germany against France in World War II). In a war testing the sides' stamina, the force applied may be against national assets in an effort to affect the will of the enemy's political echelon to fight (the United States against Serbia), or in an attempt to inflict casualties among the enemy forces and cause the collapse of public support for it in the continuation of the war (the Vietcong against the United States). In a war seeking to test the sides' resources, one may attack the enemy's war resources (the air campaign against German industry in World War II), or its supply lines (the German U-boat campaign in the Atlantic Ocean during both World Wars), and others. This selection is also contextual and circumstance-dependent.

Success in directing the war towards the nature of a war best suited to one's purposes cannot be taken for granted, and instead must be dealt with in a serious and intensive manner. Indeed, in World War II Germany sought to test military effectiveness, but in the end, the war hinged on the sides' national resources, the rate the resources were mobilized, and the sides' stamina. In the Vietnam War, the United States sought to test the depth of national resources and the military effectiveness of the sides, but the war hinged on the sides' civilian–political stamina. In the Yom Kippur War, Israel sought to test the military effectiveness of the sides, but the war hinged on the question of stamina and the ability to enlist the international system in order to shape the political end state. Therefore, an underlying principle is the need to understand and win the war on war's nature and on the issues the war is going to test.

It is only then, as a direct result of characterizing the end state, the issues the war will hinge on, and the main force application thread, that it is possible to define what elements of decision are sought. Then the operational level force utilization theme that will produce the decision must be defined, and only then is it possible to break this down into specific operational plans. Of course, it is necessary to ascertain whether the operational and logistical capabilities enable the realization of the war plans.

At the same time that the military effort is characterized, it is also necessary to characterize the operation of the other branches of the government, in order to create a coherent, long term, and effective grand strategy. At the very least, it is necessary to operate the military and the diplomatic efforts synergistically (on both the inter-governmental axis and the government-public axis), in an effort to shape the war narrative and the perception of its legitimacy, and at times also in an economic-industrial effort. Today, there is no agency in Israel capable of handling this multi-disciplinary process and capable of creating coherency between the branches of the government and its various functionaries. Furthermore, in cases such as Hizbollah, when the war is expected ahead of time and the stance of the international community is known and supportive, it is possible and necessary to deal with the questions well before the war, so that when it does erupt, the grand strategy is already formulated, the international ground is ready for action, and the end state is agreed upon with the relevant political allies. This too was not done by Israel.

In complete contrast to this orderly approach, Israel started the Second Lebanon War with no more than an aerial attack plan relevant for two to three days, without defining for itself ahead of time the nature of the war, the desired force application thread, the desired end states and decision components, and the centers of gravity to be attacked. The list of targets was not put together in an attempt to create a coherent war rationale or to understand the rationale that emerged from the proposed target list. Even the expected (limited) achievement from attacking these targets was not clarified by the leaders. What is worse, although the political echelon approved only a very limited application of force in accordance with the main idea proposed by the IDF – placing the Lebanese government in the center of the war and attacking Lebanese national infrastructures – nonetheless, neither the political nor the military echelons saw fit to delay the war in order to search for orderly, alternate ideas (even the aforementioned main idea, when acted on in isolation, serves a kind of attrition/stamina rationale rather than decision). Israel did not sketch out a clear line connecting the critical points: it vaguely defined ambitious, reality-changing objectives suitable perhaps to a war, but adopted a mode of action of a limited operation, and opened fire after preparations and on a timetable suitable to a limited retaliation.

Israel did not address the most elementary question that must be dealt with before embarking on a war in Lebanon: how is Hizbollah's war paradigm to be defeated (as Egypt defeated the Israeli paradigm in 1973, Germany defeated the French paradigm in 1940, and Scipio defeated Hannibal's paradigm in the Second Punic War)?

It is necessary to investigate the techno-tactical failures of the standoff firepower-based campaign, as well as the failures of the tactical fielded echelons: the deterioration of maneuvering capability, the lack of training of fighters and commanders, the logistical shortcomings, the deficiencies in the values of the attacking echelons such as persistence in the mission, aggressiveness, and decisiveness. However, this is not enough. Next time, if Israel only aims to attack the same list of targets with low trajectory ground fire, and at the same time successfully make use of tactical storming by a trained force backed up by well-oiled logistics, then Israel has not learned enough from the failures of the war.

On a side note, another point where the Israeli analysis of the war requires more thought is the addition of a home front defensive pillar to the defense doctrine. Israel concluded from this war that it must invest resources and efforts in defense against the rocket systems (principally interception). However, the direct physical damage done by the rockets was not critical; the main harm was in their strategic effect of disrupting normal civilian life in northern Israel. Because no defensive system is going to supply 100 percent interception, and even in the most optimistic scenario of intercepting a significant percentage of rockets it will still be necessary for civilians to go into shelters or evacuate the war zones, the strategic value of a defensive system is not clear. Israel will invest resources and efforts in a defensive system whose outcome can be measured in techno-tactical/physical terms (the interception of some percentage of rockets), yet despite that system, the enemy's rocket system will fulfill its strategic objective, namely, disrupting normal civilian life in Israel and undermining the trust and cohesion between the government, the military, and the civilians. At the same time, the tremendous efforts and resources invested in establishing a defensive system and in its operation will be denied to the Israeli effort to attain the main effort of decision.

Moreover, the enemy is capable of responding in several ways to an interception system, including saturated fire and increased salvos, as well as technological countermeasures of one kind or another. This would require Israel to increase the density of the defensive system and elicit more of the same enemy responses, while the cost of the expanded rocket threat is only a tiny fraction of the cost and complexity of expanding the interception response. In other words, Israel is liable to find itself in an arms race of rocket-versus-interception system, in which each of Israel's steps costs much more and is technologically far more complex than the enemy's countermeasure.

Since there is no essential difference in the higher levels of war between 250 rocket attacks per day and 100, or for argument's sake, between 2,000 rocket attacks per day and 600, the investment in direct response against the rockets is unreasonable. The immense effort and tremendous interception resources that would be invested in order to reduce the number of rocket hits by 70 or 80 percent would be virtually meaningless in terms of the strategic effect they would generate. It is better for Israel to invest its resources in attaining a rapid decision against the enemy and thus reduce the amount of time its home front is exposed to the threat, instead of allocating vast resources to a defensive effort lacking significant strategic value, when those resources will be missed in the main decision effort.

## Operation Cast Lead: The Next Test Case

On December 27, 2008, Israel embarked on Operation Cast Lead in the Gaza Strip. At the tactical level, the air force squadrons and task organized brigades that operated in the Gaza Strip carried out their missions virtually flawlessly within the given timetable, demonstrating clear superiority and full control of the arena. At the techno-tactical level too, the excellence of the IDF was fully apparent as the IDF provided intelligence gathering and aerial fire assets to relatively low ranking ground forces, and applying airpower even in close support missions under challenging weather conditions.

At the same time, Israel repeated some of the mistakes of 2006 in terms of managing the campaign at the higher levels. In particular, it was clear that there was some difficulty in formulating explicitly and at the right time what the desired political end state should be; what the consequent military end state should be; and how these end states would indicate a center of gravity to attack and define a comprehensive continuous campaign theme for the application of force.

The objective of the operation was defined as delivering a severe blow to Hamas in order to create the conditions for improved security arrangements for Israel. However, both parts of that sentence are vague and cannot serve as a sufficient basis for planning and executing a campaign. It is not at all clear what constitutes "a severe blow to Hamas"; some claimed this was achieved with the opening air strike, while others maintained that it was not achieved at all in the course of the operation. Worse,

**Table 6.3** Potential Objectives of Operation Cast Lead

| Campaign objective | Deterrence | Agreement with Hamas | Agreement with Egypt | Change in physical reality |
|---|---|---|---|---|
| Definition of the objective | Deterring Hamas from continuing rocket fire against Israel | Arriving at a ceasefire agreement and preventing arms smuggling | Agreement on a mechanism to stop arms smuggling on Egyptian soil with international cooperation | Unilateral, physical halt to the smuggling of weapons |
| Campaign theme | Intensive assault on Hamas' military, government, and financial assets | Gradually escalating pressure on Hamas, ending with a credible threat to overthrow its government | Demonstrating the situation's lack of stability to the international community and applying indirect levers on Egypt | Permanent occupation of the Philadelphi axis |
| Duration of the campaign | Short | Extended | Extended | Short stage of intensive fighting changing to low intensity stabilization |
| End mechanism | Unilateral cessation of the campaign | Leaving the Gaza Strip after an agreement with Hamas | Unilateral exit (vis-à-vis Hamas) after an agreement with Egypt | Fighting ebbs and replaced with permanent maintenance of the area |
| Importance of the diplomatic channel | Low | High | High | Negative (need to neutralize international involvement) |
| Reliability and stability of the end state | Medium-low | Low | Medium-low | High |

"creating conditions for improved security arrangements" is an obscure goal and may include a wide range of end states, campaign themes, and concepts that differ from one another in essential ways. Different interpretations of the directive generate fundamentally different ways of applying military force.

Indeed, "improved security arrangements" can be translated at the very least into five different end states and exit mechanisms: first, creating a reality based on deterring Hamas without any kind of agreement; second, a ceasefire agreement with Hamas; third, an agreement with Egypt and international players to set up mechanisms on Egyptian soil to prevent weapons smuggling into Gaza; fourth, permanent Israeli occupation of some Palestinian areas (the Philadelphi axis) that change the physical reality in terms of arms smuggling into the Gaza Strip; and fifth, occupying the Gaza Strip and destroying terrorist infrastructures while continuously preserving Israel's freedom of military operation in the Gaza Strip (similar to the West Bank after Operation Defensive Shield in 2002), possibly in conjunction with an attempt to allow Fatah to reassume governance of the Gaza Strip. Each of these "arrangements" engenders different end state mechanisms, exit strategies, and campaign themes, and each entails a different weight and direction for the parallel political effort. Table 6.3 analyzes four of the alternatives.

Taking ground maneuver as an example, it is obvious that maneuver is fundamentally different with each of the alternatives. If the objective is deterrence, the ground maneuver must take on the features of a large scale raid, and would not aim toward a stabilization line or entail the transition to a static defensive stage; rather, it would involve a relatively quick unilateral in-and-out. If the objective is to arrive at a ceasefire agreement with Hamas, then the ground maneuver must be more like a gradually tightening siege of the governmental center of gravity in the city of Gaza. If the objective is to create indirect leverage with regard to Egypt, the ground maneuver could be used, for example, to create a flow of refugees heading for the Gaza-Egypt border crossing. When the objective is taking the Philadelphi area, the ground maneuver must perforce be directed towards that same region. The same goes for the selected center of gravity: it would change accordingly – the assets of Hamas' military wing, or the governmental center in the city of Gaza, or the Philadelphi axis, and so on.

The inter-echelon dialogue did not sufficiently define these issues at the appropriate time, and in hindsight the campaign may be described as having applied "overall" pressure to exhaust Hamas, a campaign that ended with an arrangement with Egypt about preventing arms smuggling in its territory with a certain amount of internationalization of this arrangement (or rather without any direct connection to formulating the pertinent military moves). In addition, the direct physical blow to Hamas' military wing was limited, though Hamas did lose its political and military will to continue the confrontation, if only temporarily.

The lack of inter-echelon definition of the campaign's objectives and the desired exit mechanism led to the creation of two extended and pointless delays first, between the opening aerial attack and the beginning of the ground maneuver, and second, after the ground maneuver had reached the stabilization line. In fact, the delays lasted longer than the time spent on concerted and decisive action. These delays do not merely cast a negative light on the effectiveness of the campaign at the operational and the tactical levels, but also reveal weaknesses at the strategy and grand strategy levels, including raising the risk of a second front opening (in southern Lebanon, for example), raising the threat to the stability of friendly Arab regimes, allowing criticism of the operation among international public opinion to gain strength, and raising the risk of international involvement that would thwart Israel's attempts to realize the operation's objectives. Moreover, as emerged also in 2006, Israel's allies are prepared to accept international criticism and supply Israel with the time required to fight only as long as Israel is seen as knowing what it is doing. However, when Israel is seen as indecisive or lacking purpose, its allies are not enthusiastic about continuing to pay the diplomatic cost for supporting Israel, and the political window of opportunity narrows.

The delays described above raise yet another question, this one concerning the timing: Israel embarked on the operation on December 27, in other words, when international diplomatic systems were all but suspended for about ten days due to the Christmas and New Year holidays. Embarking on the operation at this time was consistent with the desire to neutralize the diplomatic system and seize a window of opportunity for intensive, quick, continuous military action in order to amass "hard" achievements such as, for example, occupying the Philadelphi area. However, surprisingly, an entire week of this window of opportunity was wasted on the unclear delay between the opening aerial assault and the beginning of the ground maneuver. If Israel sought a window of opportunity for intensive military action, it is uncertain why it waited. And if it sought to operate militarily in a limited way and only apply indirect levers on the international system, it is not clear why it embarked on the campaign at a time when diplomacy was all but paralyzed. The dimension of time certainly did not receive the attention merited.

In any case, Israel's dialogue with its allies about the exit mechanism could and should have taken place long before embarking on the campaign rather than in real time, particularly in light of the not inconsiderable difficulties stemming from communications problems, the pressure of the developments, and international public opinion. Indeed, the lack of prior dialogue resulted in the confusion surrounding the United States vote on Security Council Resolution 1860. Moreover, if the main axis of the campaign was the diplomatic process and the function of the IDF was merely to create the appropriate context for the conduct of that process, then Resolution 1860 demonstrates that the diplomatic channel was definitely managed in an improper manner. Even the fact that starting January 15 the political level publicly revealed its lack of enthusiasm for expanding the campaign and by its own action solidified Hamas' stance shows the problematics of the political conduct.

The problems of the inter-level dialogue, and especially the obscurity of the political level's directive to the IDF, contributed to the fact that ultimately the political echelon was the one to choose the operational plans for execution: the political echelon decided which force would maneuver as well as where and when. Operational decision making of this kind by the political level significantly neutralized both the function of the senior military headquarters and orderly military staff work. Indeed, from a military perspective it is hard to understand why the ground maneuver was formulated the way it was, why a stabilization line was chosen that lacked clear military logic and rather had clear military disadvantages, and the military reasons for the extended delay at the stabilization line. Worse, it is hard to argue that the maneuver – in the way it was designed – was in fact the most effective military action to achieve the objectives of the campaign.

At a higher level and not as the result of orderly staff work, Israel found itself designing a campaign that included relatively long durations, exhaustion and attrition of Hamas (along with exposure to Hamas rocket counter-fire), and an internationalization of the confrontation and the exit mechanism. Israel was lucky that in Operation Cast Lead it confronted the weakest of its enemies, and perhaps its least capable one. It confronted an enemy suffering from international isolation, and despite the rhetoric, from inter-Arab isolation as well. Even the Palestinians in the West Bank failed to enlist in Hamas' cause. Therefore, in the circumstances of this particular campaign, one may claim that Israel succeeded in its effort at exhausting the enemy's will to fight over time and the internationalization of the exist mechanism.

Nonetheless, three warnings are appropriate here. First, Israel must avoid repeating this pattern of action against stronger, more sophisticated, and cold-minded enemies, or enemies that enjoy wider international support. Second, conducting a campaign in a modular, rolling fashion entails long periods of fighting (and correspondingly long exposure of the civilian rear), depends on good control of the international system (to prevent the cessation of the campaign before the objectives have been attained), and requires the allocation of sufficient assets also as reserves for additional fronts. Therefore, this is not well suited to Israel's strengths and weaknesses. Third, the reliability and stability of end states that depend on subduing the enemy's will to fight and on the internationalization of end states are not high. The physical blow to Hamas' military wing was limited, and Hamas lost primarily its will to fight. But the will to fight is a fickle phenomenon, and changes according to circumstance, context, and time. This opens the door to repeated future attempts to test Israel's reaction threshold, to erode the status quo, and gradually to wear away the campaign's achievements. As for the international stabilization mechanisms, the difficulties of enforcement are well known from similar situations, and a reference to Security Council Resolution 1701, which is not actually enforced, is sufficient to make that point.

As for Hamas, it failed in shaping the confrontation at practically every level. Hamas' effectiveness in its attempt to harness the international community on its behalf was low, and its allies did not affect the formulation of the confrontation's exit mechanism or diplomatic envelope. Hamas thought to adopt Hizbollah's strategy and operational concept, but failed to see the fundamental difference between the two cases. First, it failed to understand the significance of the vast geographical, topographical, and political differences. Second, Hamas did not internalize that it is not a non-state guerilla organization as Hizbollah was in 2006 (though it is not clear that Hizbollah is still such today), but rather a de facto state. Hamas has vulnerable governmental assets, it bears responsibility for the population of the Gaza Strip, and it has what to lose – governmental control. As such, it does not have the freedom of action of a true guerilla organization and it does not have the ability to truly fade and disappear. In fact, the most important element for Hamas in the course of Operation Cast Lead was not the rocket fire at Israel or resisting the ground maneuver, but the attempt to maintain continuity in demonstrating the presence of civilian police personnel in uniform in the Gaza Strip cities and towns. Even Israel was to a certain extent surprised by the fact that Hamas' principal center of gravity was state-like rather than military, and thus the planning of the campaign did not reflect an adequate attempt to exert pressure on this center of gravity or interfere with the process of Hamas' governmental recovery from the initial aerial assault.

At the field levels, Hamas found it difficult to preserve continuous rocket fire, which thus ebbed as time passed. Hamas also showed very low tactical capabilities when confronting IDF maneuvering. Furthermore, Hamas failed in its attempt to turn obstacles, booby traps, and mines into the main response to the maneuver.

From a wider perspective, Operation Cast Lead was an important step in Israel's confrontation with the new paradigm of fighting Hamas, Hizbollah, Syria and others. This paradigm is based on three principles: first, the primary form of the enemy's fighting is attacking the Israeli civilian front with rockets and missiles. Second, to the extent possible the enemy avoids direct military confrontation in large scale battles between its fighters and the IDF. Third, the enemy hides among and behind its own civilian population and uses it as human shields. The vital link allowing the Syrian–Hizbollah–Hamas paradigm to exist is that the enemy "appointed" Israel to serve as the guardian responsible for the safety of the enemy's civilians, while Israel in the past tended to accept that responsibility and therefore acted with a great deal of restraint. However, shouldering the responsibility for our own civilians while accepting responsibility also for the enemy's civilians, thereby releasing the enemy from that responsibility, created impossible, lose-lose situations. In Operation Cast Lead, the IDF operated more freely than in the past – though still within the limits of international law – against the enemy wherever it might have been hiding, even among its civilians. Indeed, attacking the enemy's combatants and weapons wherever they were, even in the basements of mosques, public buildings, and residential quarters, represents a measure of breaking the enemy's paradigm.

In addition, the campaign contributed to familiarization of the civilian systems in Israel with rocket and missile fire, and the effect generated by the enemy was weakened in comparison with the confrontation with Hizbollah in 2006. The political and public systems did not collapse even when 1,300 Israeli homes were damaged, Israel was not deterred by Hamas' rocket fire, the rocket fire did not hamper the freedom of action Israel assumed, and it did not represent a serious constraint in Israel's considerations.

Finally, it is possible to undermine the war paradigm of Syria, Hizbollah and Hamas in alternate ways as well, and it would be dangerous to draw far reaching conclusions on the basis of Operation Cast Lead both with regard to the desired method for undermining the enemy's paradigm and with regard to the reaction of Israel's civilian rear if attacked by fire of greater capacity and precision as might be expected from a confrontation with stronger enemies.

# 7

# The Future War

## Parallel War against a State Enemy that has Adjusted to Fighting against RMA and Adopted a Guerilla Paradigm

This chapter presents a possible model for the next generation of warfare that Israel and similar nations may face: war against a state enemy that has adjusted to fighting against RMA capabilities by adopting the guerilla paradigm (similar to that of Hizbollah). Analytical tools and concepts presented in the previous chapters can shed light on the future war. The chapter also makes the following claims:

1. Deeming the 1991 and 2003 wars in Iraq as models for the war of the future against a state enemy is liable to be a misleading exercise, because the Iraqi military never adjusted to contending with American capabilities and never presented an alternate war paradigm against the American paradigm.
2. Nonetheless, RMA's rapid target destruction capabilities and Hizbollah's successful experience against the IDF in 2006 are liable to impel state enemies to adopt the guerilla paradigm.
3. A regular military that adopts a guerilla paradigm will on the one hand try to attack the civilian–political stamina by creating a situation that is intolerable over time using strategic firepower or attrition in the theater of operations, and will on the other hand hide physical centers of gravity, lower its signature, decentralize its mass, and not assume positions for the prime major battle.
4. Under such circumstances, the utility of the blow directed against the enemy's forward fielded formation is reduced. There is thus no choice but to attack the enemy's strategic centers of gravity – but only such that serve a nature of decision rather than those involving questions of stamina and attrition.
5. Maneuver is a key tool in shaping the nature of a war of decision. Maneuver that creates an encounter between armed forces hastens and intensifies the test of the sides' military effectiveness, and the

rapid maneuver towards a strategic center of gravity denies the enemy the freedom of action to continue with an extended war of attrition. The correct maneuver forces the enemy to react, and that reaction is likely to undermine the enemy's guerilla paradigm.
6. Maneuver is capable of creating enemy operational centers of gravity unintended on the part of the enemy, and of enabling us to force a prime major battle on our own terms. The maneuver that creates a concentration of enemy mass also supplies the optimal conditions for operating firepower effectively.
7. However, a tactical maneuver may in some circumstances end with operational and strategic-level exhaustion. A frontal maneuver into a saturated defensive formation, or one that overstretches us logistically, is liable to actually test our stamina. It is necessary to shape the maneuver properly so that it tests the sides' military effectiveness at every level of war, and does not mistakenly create a test of stamina.
8. Achieving decision against a state enemy that has adopted a guerilla paradigm is not possible in a relatively restrained operation, unlike a classical restrained decision of the enemy's military mass that deploys with a high signature for the prime major battle.

## The Operational and Strategic Level Significance of a State with a Guerilla Paradigm

After a series of examples, circumstances, and types of war that add a new perspective to classical Clausewitzian doctrine and illustrate that it is not sufficient for addressing the complexity of circumstances such as those prevailing in the Second Lebanon War, it is time to assess the next generation of warfare: the war of the future against a state enemy that has adopted the guerilla paradigm. This type of warfare is much more complex and requires the construction of appropriate analytical tools. Concepts and challenges analyzed in previous chapters are important here, for example, the war over the nature of the war, i.e., is this a war of decision or a war of attrition; inter-state asymmetry; the parallel war in which each side conducts a separate campaign against its opponent; the looser link to the war's outcome at the different levels; the guerilla paradigm; and Hizbollah's approach of unflagging defiance – based on a force composed of many similar components that continues to operate effectively, even if many of the components are lost; and "winning by not losing" or outlasting the democratic free market foe.

Indeed, the race for techno-tactical advancement between Western countries that adopted RMA or some of its elements and some of the nations that represent potential enemies of the West is already in high gear, and one can clearly distinguish the acquisition of means to disrupt intelligence gathering capabilities, the operation of precision weapons, and wireless command and control used by the West, as well as its freedom of flight. However, not enough thought has been given yet to understanding the operational and strategic level significance of an asymmetrical confrontation against a regular state military that has adjusted and found appropriate responses to Israeli and Western military intelligence gathering capabilities and aerial attack capacities.

Worse still, viewing the two recent wars in Iraq (1991 and 2003) as a model of the future war is liable to be misleading, because the Iraqi armed forces never adjusted to contending with the American capabilities and operated in a mode that complied with the American paradigm and war plans. In both cases the Iraqis deployed large, outdated military frameworks with high signatures in an exposed desert area in order to try to confront symmetrically the world's most advanced, well-equipped, and organized military. Moreover, at times the Iraqis moved large military formations over long distances, on open ground, in broad daylight, and under clear skies indisputably dominated by the US. The Iraqis volunteered as it were to place themselves in the heart of the American efficiency envelope, and actually played the exact role the Americans had scripted for them. The 1991 and 2003 wars are instructive mainly with regard to the Iraqis' lack of understanding of the sides' strengths and weaknesses, their lack of ability to adjust to a new reality, and their inadequacy in shaping a paradigm that would challenge the American approach. However today, and all the more so in the future, various state forces' adjustment to the West's intelligence gathering capabilities and precision firepower may well push some of them to operate according to a logic, nature, and mode similar to those that were identified more with guerilla organizations than with regular forces. The future enemy dealt with in this chapter is thus the state military operating to a large extent according to a guerilla paradigm.

The American RMA and its related technologies have some clear advantages, such as the capability for rapid destruction of large military formations with a high signature in open terrain, and at least at this stage, potential enemies are still hard pressed to provide direct techno-tactical responses to them. Therefore the capability to destroy armored or mechanized enemy forces engaging in classical maneuvers is excellent. After all, RMA's technologies were born from a search by the United States and NATO for ways to break a massive Soviet attack in Europe.

Therefore, the offensive strategy of the state enemy that has adjusted to war against RMA and precision firepower must be asymmetrical and go beyond the classical mode of war. Strategy will not be based on an attempt to realize the war's objectives in a direct manner, but rather indirectly, by creating a situation that on the one hand is exhausting and intolerable over time for the Western democracy, and on the other hand, offsets the advantages of Western intelligence gathering and firepower and in fact attempts to deny us the opportunity for a military decision and a military exit from the war. This is somewhat reminiscent of the Egyptian attempt in 1973 to neutralize the IDF's capability to achieve a military decision, and to extend the war as it saw fit in such a way as to leave only a diplomatic exit from the fighting, which required Egypt's agreement. In this mutually accepted exit, Egypt sought to advance the process towards attaining its war objectives.

Such a strategy primarily attacks the stamina of Israel's civilian–political system or that of any Western democracy. As with Hizbollah, such an attack is possible in two ways: indirectly, by causing casualties to the soldiers in the battlefield, and directly, by attacking the civilian home front, using rockets and missiles.[60] This enemy's blows to fielded formations are not meant to achieve a decision or a classical operational objective such as territorial conquest, rather to impact on the Western democracy's willingness to fight by undermining the trust and cohesion between the government, the civilians, and the military. The blow to the will to fight requires a number of conditions, among them a sufficiently extended period of fighting, and the absence of any sense among the civilian population that a concrete achievement justifying the cost of the war still exists: every additional day of conflict is another day without results, the war is seen as pointless, and as such it must be stopped sooner rather than later. Thus gradually, the recognition is liable to form that it is best to extricate oneself from the war at the earliest opportunity, even at the cost of a mutual agreement that includes giving the enemy a political achievement or a unilateral withdrawal from the challenge.

Attacking the home front directly with rockets and missiles obligates the enemy to develop a force that will survive and outlast RMA's intelligence gathering and precision strike capabilities: starting from operating outside the intelligence gathering envelope (under the signature threshold, at a fleeting pace, outside the discovery range, outside the weather envelope, by simultaneous saturation of sensors, and so on), and ending with very high redundancy and stamina that will enable the rocket force to absorb extensive and effective blows and still succeed in generating enough defiance firepower over a long period of time.

Another fundamental component in the said enemy's future war strategy is the attempt to deny Israel or the Western state the opportunity for a military decision and a military exit from the war. This may be achieved in a variety of ways: because the enemy strategy is based on an indirect approach and attrition, the enemy is capable of giving up heavy, maneuvering formations and base itself on saturating the battlefield with light units operating with a low signature attempting to achieve the highest possible kill rate. In Israel, there has recently been much talk of the "empty battlefield." On the contrary, the battlefield will not be empty; it will be fully saturated by an enemy with great firepower but a low signature. Neutralizing RMA and its attempts to strike at the enemy's functioning as a system (EBO) can also be achieved through the buildup of force and adopting a mode of "an organization not functioning as a system." Like Hizbollah in 2006, a regular military is also capable of operating as a flat, decentralized network of territorial autonomous fighting cells. The fighters, weapons, and logistics are deployed in every territorial cell ahead of time, most of the orders are preplanned and determined in advance, and there is very little need for transporting forces and supplies or for depending on complex command and control systems. The organization is in any case constructed in such a way that it is not expected to operate like a system with active and intensive interaction.

The transition from a classical military structure to a network of autonomous, decentralized cells is significant in many other ways as well. Its prominent disadvantage is that such a network of cells would have a difficult time responding with systemic effectiveness to a maneuver through it: it would lose its rationale and advantages should it try to amass forces for a defensive effort, not to mention the execution of a counter-offensive of a scope exceeding the limitations of the autonomous tactical cell. It would also find it difficult to effectively utilize assets belonging to the higher command levels, such as intelligence gathering or mobile reserves. In practice, each cell would deal almost on its own with inflicting the most casualties on the force maneuvering through its territory, until it collapses. Clearly, such a network would also lose its rationale and advantages were it to try to maneuver into an unexpected area that was not preplanned (maneuvering that requires amassing force, a high signature, and operating a logistical system), and therefore bypassing it would create difficult dilemmas for the enemy.

At the same time, such a force has many advantages. A network of autonomous cells does not have a clear operational center of gravity that if struck would bring about the collapse of the rest of the system's elements or the dissolution of its original structure. The cell network structure requires fighting separately against each cell located in our area of operation, and it is difficult to find weaknesses, concentrations of forces, or nodes that if attacked would impede the enemy's ability to continue operating effectively. Each cell operates separately in the main, with limited inter-cell relationships and functioning, and therefore it is difficult to strike directly at the enemy system in a way that would disrupt its systemic-operational order to the point where it would be unable to fulfill its originally designated function. A blow against a network of cells at best resembles a shot in the dark, and at worst a blow to a punching bag filled with nails: it will definitely hurt, but will not crush the bag. For these reasons, there is also a difficulty in conducting the major battle of decision.

# THE FUTURE WAR

The thinning out and obscuring of the enemy's physical centers of gravity to the point of their disappearing diminish both the anticipated achievement and anticipated usefulness (at the upper levels of war) of a blow directed at an enemy's forward fielded formation.

An additional important feature in the next generation of warfare is that in order to attack our stamina over a long period of time the enemy does not have to maintain full classical military effectiveness. It is sufficient if the enemy has enough residual capability to continue its defiance. Hizbollah may be seen as maintaining defiant firepower whether it launches 300, 200, or "only" 100 rockets a day at Israel, and the Vietcong maintained defiant guerilla capability against the United States whether in any particular battle it killed 100 American soldiers or ten. Even an outstanding tactical achievement, such as reducing the number of Hizbollah launches from 250 rockets per day to 100 (had this occurred) would not have been enough to deny Hizbollah its capacity to generate the strategic effect it sought. Therefore decision, i.e., denying the enemy force its ability to operate effectively in order to fulfill its mission, cannot under these circumstances be measured in tactical or physical terms, but only in terms of denying the operational level or strategic achievement that the enemy seeks to realize. The desire to directly deny the enemy its residual defiance capability (for example, reducing the number of rocket launches to a negligible number, which is difficult to define in advance) is an ambitious benchmark, in many cases too far reaching to be attained. For this reason there is also a real difficulty in attaining a classical military-operational decision.

A consequent additional difficulty relates to ending the war. Historically, the military-operational decision usually led to an end to the war, which occurred soon after the military decision was achieved. However, the difficulty that arises in an attempt to cause the enemy to lose its residual defiance capability challenges our capability of ending the war, even if the operational plans' objectives are attained. Even maneuver to the preplanned line and destruction of the enemy's mass on a significant scale, which would impair (but not eliminate) the enemy's tactical capability, are liable to not deny the enemy its capability to remain defiant over time at the operational and strategic levels.

In such a case, the war of the future turns into a parallel war: not an encounter between two masses on the same battlefield in a symmetrical attempt to destroy one another, rather an action of each side with different parameters, with the enemy trying to avoid the major battle of decision and operating both directly and indirectly to exhaust our civilian system. In such a situation, it may well be possible that the two sides will realize their operational plans without their contradicting or disrupting one another, which further forces the question of what constitutes decision. In realizing the plan, the enemy enjoys an additional asymmetrical advantage: even if the enemy requires time in order to erode our civilian–political system of the will to fight, in certain senses the enemy begins the war already holding some parts of its desired end state: from the outset, the enemy is in a not-losing position (striving only for "winning by not losing") and has defiance capability. In other words, there is asymmetry in the fact that at the war's starting point we are only just beginning the challenging journey towards realizing our end states, whereas the enemy has already attained some of them and we have to work at depriving the enemy of them.

A state enemy that has adopted the Hizbollah paradigm thus also challenges the classical Clausewitzian doctrine:

1. The enemy is liable to win the war simply by not losing: a situation of attrition having no military exit will be created, which will cause the loss of our civilian–political will to fight, even without a military decision in favor of the enemy.
2. The enemy is capable of absorbing severe tactical blows but can continue to maintain residual defiance capability, which will enable it to realize its plans at the operational and strategic levels, and therefore we will find it difficult to attain a decision in the classical sense.
3. The said enemy will obscure and thin out its military-physical centers of gravity, and it might not even have a clear military-operational center of gravity in the war's theater of operations.
4. In the absence of a clear military-operational center of gravity and because the mass is decentralized and disappears, we are denied the opportunity to conduct the prime major battle.

## Deconstructing a State with a Guerilla Paradigm: Undermining Basic Assumptions

In that case, how does one contend with the regular state enemy that has adopted a guerilla paradigm? Hiding the operational-physical centers of gravity creates a situation that perforce – though not out of doctrinal preference – leaves mostly the option of a threat against strategic and abstract centers of gravity. Tackling some of these centers of gravity is a complex matter. First, such centers of gravity are always context-dependent, and there is no checklist of centers of gravity. Second, it is difficult to define abstract centers of gravity. Some are plans, paradigms, and basic assumptions of the enemy; how to identify them and devise a way to attack them is not simple. Third, it is critical to choose strategic centers of gravity that if attacked would shape a war designed to test the military effectiveness of the sides, and may lead to a decision within a short period of time, rather than be carried away by mistake into a war of mutual, prolonged strategic attacks, testing stamina, staying power, or the ability to enlist the international community. There is considerable temptation to act against valuable, easy-to-attack national targets, such as the Americans did in Kosovo, but as analyzed in Chapter 1, focusing on attacking them is liable to channel the war to realms that are suitable for the United States but not a small nation such as Israel.

One may categorize the relevant strategic centers of gravity whose attack is likely to be useful according to several main axes. First, there are physical centers of gravity whose attack would impair the enemy military's strategic ability to act: such centers of gravity may include enemy strategic fighting echelons, strategic reserves, forces defending the regime, command and control systems, and any other system that is critical for the enemy's maintaining strategic ability to act and strategic level posture. Examples of attacks on such systems include the destruction of the Egyptian air force at the beginning of the Six Day War, the blow inflicted on the Iraqi command and control systems at the start of both the 1991 and 2003 wars, and the destruction of the French navy by Admiral Nelson, which trapped the French forces in Egypt and isolated them from the European theater.

Second are physical centers of gravity that if attacked or threatened would impair the enemy's strategic freedom of action to continue fighting. An example is Scipio's threat against the capital city of Carthage, which forced Hannibal to abandon the campaign in Italy. In such a case, there is a sort of indirect decision, because the enemy does not directly lose its ability to act but only the window of opportunity needed to fight the war it is interested in. It is important to make the distinction that Scipio's threat against Carthage was its immediate, physical conquest, and that Hannibal had no alternative but to abandon the Italian campaign. In contrast, American aerial attacks on Hanoi during the Vietnam War raised the cost of the war but did not create an immediate, physical threat to force the North Vietnamese to stop fighting. Rather, it left in the hands of the North's leadership the option of deciding whether to pay that price and nevertheless continue fighting.

Third are abstract centers of gravity that if attacked undermine the enemy's war paradigm. Examples of denying the enemy the opportunity to fight the war it had planned and forcing the enemy to fight a different one are the Egyptian attack on Israel's war plans in 1973, and the German attack on the French war plans in 1940.

One of the ways of undermining the paradigm of a state enemy that has adopted guerilla modes is through strategic maneuver. The concept of the decentralized network of autonomous low signature fighting cells is based on a number of assumptions that the enemy "takes for granted," among them the assumption that the boundaries of the theater of operations are predictable, that the enemy is pre-deployed in the area of operation before the outbreak of fighting, and that in every engagement the enemy will be in a position of relatively static defense. A network of autonomous fighting cells would find it difficult to redeploy in new and unexpected area of responsibility, and were it to try to transport large forces to new territory, it would lose its advantages: transporting large forces over distances cannot be done while maintaining a low signature and by autonomous tactical action. Optimal low signature is achieved with static defensive dispersal in a given territorial cell, whereas attack and movement require concentrating forces and operating a complex logistics system. Hence, a strategic maneuver that opens a new theater where the enemy is not deployed or at least expands the existing theater in an unexpected way is liable to replace the dilemma the enemy imposed on us with a different dilemma that we impose on the enemy. The enemy either enables us the freedom of action on the ground to maneuver towards a strategic center of gravity through terrain where the enemy is not deployed, or it attempts to counter-maneuver with large forces and over distances, losing the advantages of low signature and autonomous action and entering the heart of our efficiency envelope. In such a situation, any choice the enemy makes is good for us.

Indeed, the maneuver is sometimes an efficient tool for creating centers of enemy mass. For example, the German maneuver in World War II into Holland and Belgium created a center of mass of French and British forces in northern France and in Belgium. The German maneuver towards strategic centers of gravity in the Soviet Union (Stalingrad, Moscow, oil in the Caucasus) forced a transition on the Russians from withdrawal and delaying operations to a rigid fixed defense and waging large battles. A maneuver towards a strategic center of gravity may thus force the enemy to concentrate large forces with a high signature to defend that strategic center of gravity, thereby leading to the creation of an operational center of gravity and the opportunity to conduct a prime major battle – an opportunity the enemy sought to deny us. Herein lies the difference between the true guerilla and the state military trying to act like a guerilla: the Vietcong did not deploy to defend any particular territory and usually did not try to deny the Americans temporary freedom of action on the ground wherever they desired. A state military does not enjoy this luxury.

In this way, maneuver also improves firepower effectiveness. In a static, maneuver-free war, the enemy can on the one hand reduce its signature and disappear, and on the other hand dig into hardened, fire-resistant positions. Under such circumstances, precision firepower loses its advantages. However, maneuver may force the enemy to expose itself, to raise its signature, and to operate as a massive logistical and operational system. The maneuver is what makes firepower more effective and even enables functionality-based warfare (EBO), and thus the critical synergy between firepower and maneuver is created. One of the most important lessons from the air campaign by the United States and NATO in Kosovo was that in the absence of a ground maneuver on NATO's part, the Serbs did not have to concentrate a mass of force in a defensive effort or in any other concentrated deployment. Instead, the Serbian forces scattered and disappeared, and the effectiveness of the American aerial force against the Serbian forces was therefore severely impaired and ultimately completely undercut. The conclusion was that a friendly maneuver is a critical factor causing the formation of enemy masses of force, rendering the aerial firepower effective.[61] This conclusion is consistent also with the lessons the United States learned from the 2003 Iraq War, when the American maneuver caused the Iraqis to execute a significant change in their deployment, on open ground and in broad daylight. This situation granted the US Air Force an opportunity to optimally manifest its abilities on the battlefield, and enabled it to achieve rapid destruction and a decision in that part of the campaign.[62] Thus, the maneuver is not simply an archaic techno-tactical method of bringing short range ground assets into their limited firing envelopes, as claimed by maneuver's eulogizers, rather a critical and irreplaceable element in shaping the war at all levels, including the strategic (table 7.1).

Table 7.1 Influence of Maneuver and Firepower on Military Effectiveness and the Nature of War

| | | Enemy that adjusts to fighting against RMA | |
|---|---|---|---|
| | | Static defense/fire only | Maneuver |
| Western nation | Static defense/fire only | A war of attrition, not taking full advantage of our strength. | Source for RMA, but today not a reasonable scenario. |
| | Maneuver | Maneuver into enemy defense deployments: risk of attrition, not taking full advantage of our strengths | Taking full advantage of the strengths of RMA and the weaknesses of the enemy |

Hence it is crucial to understand that the point made here of attacking the enemy's strategic centers of gravity says nothing about the desirable systemic-operational or techno-tactical mode of action. In particular, it says nothing about preferring long-range aerial attacks. A military mode of action must always be chosen in relation to the entire set of circumstances; given that the enemy's center of gravity under attack is always context-dependent, in certain contexts it is precisely the ground maneuver that will enable us to attack it more effectively. Indeed, threatening strategic centers of gravity with a maneuver is a well-known phenomenon, from Scipio's maneuver towards Carthage in the Second Punic War, to Sherman's March to the Sea in the American Civil War, the Allies' attack on the heart of the Third Reich and the conquest of Berlin in World War II, and Israel's advance towards Damascus in 1973, through the destruction of the Iraqi Republican Guard and the conquest of Baghdad by the Americans in 2003.

More important, maneuver is also an essential element in shaping the nature of the war and the issues tested. Maneuver that brings the belligerent forces together, directly or indirectly, accelerates and intensifies the test of their military effectiveness, and directs the war towards a short, climactic confrontation, more likely to end with a decision. Firepower-based operations alone typically prolong the war, and in a case such as Israel's, when the firepower is reciprocal, a window of opportunity opens for other influencing elements to enter the picture as well, such as stamina, staying power, resources, and the ability to enlist the international community. Proper maneuver thus helps to shape a war of rapid decision.

Ground maneuver towards the enemy's strategic centers of gravity creates a type of direct, acute threat that no standoff fire, intensive and precise as it may be, is capable of equaling.[63] Such a maneuver provokes a sense of stress and urgency in the enemy, and yields a visible achievement. Here too rapid strategic maneuver constitutes in effect an attack on the enemy's plans to conduct a war of prolonged exhaustion and challenge of a Western nation's stamina. In order to wear down the political–civilian system's stamina in a democracy, the enemy usually needs to establish several foundations, among them a long enough period of fighting for public support for the war to erode gradually, and our own inability to point to significant visible achievements. A quick maneuver towards the enemy's strategic centers of gravity is likely to deny the enemy both of these fundamental elements.

Moreover, in certain contexts the territory itself constitutes a strategic center of gravity, because threatening the territorial integrity of the enemy state constitutes an attack on the enemy's regime internal legitimacy, its internal prestige, and even its right to exist, and thus creates a strategic threat to it.

At the same time, in order to achieve the preferred nature of war, the maneuver must be properly designed, and thus a tactical maneuver, for example, not devised appropriately at the upper levels of war may produce precisely a war of attrition. Thus, the Wehrmacht's maneuver eastwards that pushed the Red Army into the interior of the Soviet Union (1941–43) eventually created a position of inferiority at the operational and strategic levels where the Wehrmacht suffered from overstretching across a 1,500 km-wide front, long and vulnerable lines of supply, and exhaustion, while the maneuver at the start of World War I, in a saturated and obstacle-ridden battlefield and into the enemy mass, actually produced the trench war. Therefore, the design of the maneuver must be such as to ensure a rapid test of the sides' military effectiveness at every relevant level of war, and not accidentally produce a test of other issues (such as stamina and staying power).

Beyond maneuver, other points can also work in our favor in a confrontation with the enemy described in this chapter. Reality is much more complex than the schematics of doctrine, and it is usually not possible to apply the doctrine in practice in full. An enemy operating to reduce its signature will still create a medium signature of one kind or another. An enemy acting to thin out and obscure its centers of gravity will still leave critical weaknesses of one kind or another. And even an enemy seeking to operate not as a system cannot exist in a vacuum. At the end of the day, a state enemy – as opposed to a guerilla organization – has clear command and control systems and logistical mechanisms. Usually, the military commander will see to orderly supplies for his units and will want to be able to intervene and affect the course of the fighting. Hence, he will want a mechanism through which he will receive regular information about the state of the fighting, and will strive to be equipped with the capabilities to have his decisions come into play through the chain of command, communications, and assets directly under his command. All of these require some sort of signature and function in any case more or less like a system. Therefore, along with the necessity of attacking the enemy's strategic centers of gravity, it is also necessary to take full advantage of the possible achievement of rendering a direct blow to the enemy's military capability to act and destroying the enemy's forces and fielded formations.

The approach to achieving decision in the next generation of warfare against states operating with a guerilla paradigm is apparently to be found in an optimal combination of a threat against the enemy's strategic centers of gravity, an attack on the enemy's war paradigm, and maximizing the potential for a classical attack on the enemy's capabilities and creation of a critical mass of destroying the enemy's forces. Finally, as in Israel's mode of decision in 1973, demonstrating tactical military superiority in terms of force-on-force and dominance on the battlefield, which is achieved through the destruction of the mass of the enemy's forces, has a substantial effect on attaining decision and on the projection of decision to the upper levels of war. The other side of the coin is that Israel's reluctance to engage in a direct or indirect tactical force-on-force confrontation in the South Lebanon battlefield in 2006 created a significant problem in Israel's concept of power even at the highest levels of war (and peace). Hence the tactical, direct, and "simple" destruction of the enemy has the potential to create strategic value, and this, after all, is the "annihilation principle" that Clausewitz wrote about.

Thus, in the background of the war, an asymmetrical struggle is conducted to determine its nature: the enemy tries to test the sides' stamina, by attacking the civilian home front directly, by disappearing, and by not taking positions for the large battles. Nonetheless, we still try to impose a quick war of decision, by threatening strategic centers of gravity and undermining the basic assumptions of the enemy's war plans. Such a scenario has far reaching implications; above all it is evident that in order to attain decision and victory it might be necessary to maneuver beyond the traditional theater of operations and into the strategic depth, i.e., it will no longer be possible to attain decision in a "restrained" war limited to attacking the enemy's forward fielded echelons. Even the Yom Kippur War, perhaps the most violent of the Arab–Israeli wars, was limited to a clash between military masses in a traditional theater of operations, and both sides took pains most of the time to apply force only against military-operational targets. Such a traditional, restrained outline of warfare might be replaced by mutual unrestrained attacks on strategic centers of gravity.

## The War becomes Less Restrained

Two concluding points make the picture even more complex. First, precision arms technology enables the IDF and Western armies to achieve precise and effective strikes against enemy targets, while at this stage, the enemies must make do with imprecise fire against large ground targets (such as Hizbollah's attacks on Israel's northern cities). While this type of fire induces distress and disrupts normal civilian life, its direct physical damage is limited. However, precise navigation of weapon systems is becoming simple and accessible, and sooner or later whoever pays their relatively low cost will acquire them. A future symmetry in precision strategic firepower will create an asymmetry of vulnerability between an industrialized nation and a third world country. Industrialized democracies have many more assets of infrastructure and economics than third world countries, are much more sensitive to the economic costs of the war, have civilians who are much more sensitive to the disruption of their normal routines and welfare, and governments and militaries that must be much more attentive to the morale of the civilians and their wellbeing. The enemy arming itself with precision weapons capable of impairing the effectiveness of state infrastructures therefore constitutes another turning point in the war's paradigm, and will force us to achieve a strategic surrender that is swifter, more violent, and less restrained. Moreover, if the enemy attains deep precision strike capability against our military assets, it will present a real challenge to our military effectiveness, and this challenge too may necessitate a further change in the war paradigm.

Second, Prof. Michael Handel has claimed that RMA's techno-tactical superiority is liable to push third world countries lacking the capability of symmetrical equipping into arming themselves with weapons of mass destruction: "The readiness of such states to employ crude weapons of mass destruction may quickly reduce the appetite of high-tech military powers to wage war.... This is certainly one type of circumstance in which low-tech can defeat high-tech."[64] Yet we have already concluded that in a war against a state enemy that has adopted a guerilla paradigm we would find it difficult to attain a restrained military-operational decision limited to attacking the enemy's fielded formations, and therefore we have no choice but to seek a decision against that state by threatening the enemy's strategic centers of gravity. That is, we are liable to be facing two bad choices: waging an extreme, unrestrained war in which we attack the strategic centers of gravity of an enemy armed with weapons of mass destruction, or withdrawing from the challenge because of an inability to attain a rapid decision, thereby giving the enemy an achievement (i.e., losing the war).

Although the said enemies of Israel, the United States, and the West might be pushed into adopting the guerilla paradigm out of inferiority in the test of a symmetrical clash between large military formations, and adopting the guerilla paradigm seems to present a lower level of threat than the threat of classical ground invasion, the adoption of the guerilla paradigm makes the war more complex. It is more difficult to identify operational centers of gravity, it is more difficult to conduct the prime major battle, it is more difficult to attain decision, and it is more difficult to define the military exit from the war. Therefore, the next generation of warfare might focus on mutually attacking strategic centers of gravity. Such mutual strategic attacks, when carried out against future potential enemies armed with precision weapons and with unconventional weapons, make the war of the future less restrained and more violent.

The key to decision and victory is not to be found in destroying the enemy's fighting mass alone, even if only because of the difficulty in locating that mass. The key is to be found in defeating the enemy's paradigm and plans, that is, in planning the war correctly. The days that are critical to attaining decision and victory are not necessarily when the battles are being conducted but rather when the war is being planned – or, in other words, right now.

# Conclusion
## It Never Ends

### "Rock-Paper-Scissors"

War often resembles the children's game of "rock-paper-scissors."[65] When we have rock-like features, the enemy develops paper-like qualities, and then we are forced to develop scissors-like capabilities, in response to which the enemy develops rock-like features, and so on and so forth. And just as the game "rock-paper-scissors" has no dominant move creating an absolute and permanent advantage over the opponent, so too in war: there is constant changing and shifting of weight among maneuver, firepower, disappearance, and protection, between mobility and saturating the battlefield with forces and obstacles, between the advantages of defense and offense, as well as between stamina and rapid decisive capabilities. New paradigms replace old ones, and the wheel continues to turn. RMA created a clear advantage in the area of destroying high signature targets in open terrain, but it did not succeed in creating unanswerable capabilities and it did not bring the history of war to an end.

To some extent, RMA caused the degeneration of the Western war's DNA. There is no universal correct way to wage war, and historically one may discern different approaches to war, even within the West itself. For example, until the middle of the twentieth century, Prussia/Germany saw itself as a small nation relative to its enemies, threatened simultaneously on several fronts, and lacking the resources and the stamina necessary to conduct a prolonged war. This view required Prussia/Germany to wage short, bold, dynamic, and creative wars, based on outstanding military field leadership, brilliant strategy, and risk-taking. By contrast, the United States sees itself – and justifiably so – as the most powerful nation in the world, so that even when its global interests are threatened, it is itself always safe (with the exception of nuclear warfare), and can therefore wage wars on its own terms and not on the basis of enemy dictates. The Americans have all the time they need, and in fact, three and a half years elapsed from the Japanese attack on Pearl Harbor until Japan's surrender. America's unique power enables it to conduct campaigns to exhaust the enemy's strategic core based on overwhelmingly excessive force, and the country's geographical distance from the theater of operations shields its home front from exposure. During the 78 days in which the United States

## CONCLUSION

carried out 38,000 sorties against the Milosevic regime, not a single American civilian suffered so much as a scratch. Therefore, the United States is exempt from the need to move quickly, take extra risks, and rely on bold military leadership.

With some exaggeration, one may say that the American way of war is based on the logistical-administrative-organizational mediation that brings the products of the vast United States defense industry to the enemy's leaders and soldiers. Even the American definition for the word "strategy" is the allocation of resources to fulfill national objectives (a definition that cannot be appropriate for Vietnam or Egypt, and not even for Germany or Israel). Not surprisingly, the more prominent German military leaders, such as Heinz Guderian and Erwin Rommel were bold field commanders (sometimes to the point of recklessness), whereas the military commanders of the most successful and glorious war in American history, Dwight Eisenhower and George Marshall, served significant portions of their careers in staff and administrative positions without accumulating significant battlefield experience. Even bold American field commanders such as George Patton relied on excessive tactical and logistical might and affluence of weapons, without needing to be brilliant at the operational or strategic levels. Not surprisingly, Patton is credited with the following saying: "Good tactics can save even the worst strategy. Bad tactics will destroy even the best strategy." (Presumably, Ho Chi Minh would not have agreed.)

Even Britain, relatively close to the United States in character and circumstance, differs vastly in its national way of war. Britain was a great naval power, but had a limited army and a high degree of sensitivity to attrition. Therefore, the traditional British approach to dealing with continental enemies rested on three axes: use of the British navy to create a naval blockade; money and resources to support the buildup of the forces of continental powers that were members of the same warring coalition; and participation in auspicious peripheral ground campaigns. In fact in World War II, on the basis of this approach, Britain persuaded the United States to arm the Soviet Union and embark on campaigns in North Africa, the Balkans, the Mediterranean islands, and Italy, while still remaining skeptical about "continental commitment" on the ground in western Europe. British theoreticians, headed by Liddell Hart, tried to turn the British way of war into a universal formula of preferring the indirect, measured approach, carefully weighting objectives versus cost of attainment, and meticulously choosing the challenges that should be met head on. In response, Paul Kennedy asked how exactly were nations such as Poland supposed to apply the British way of war.[66] The Americans, for their part, viewed the British approach to World War II as undesirable "scatterization," and looked for the main front and the primary mass where the American steamroller of resources, logistics, and organization would be best manifested. The Americans never saw the point of dividing their resources among several different efforts, because when each side puts all its cards on the table, the relative strength of the United States finds maximal expression.

## The American Way of War Suits Only the United States

Even though the American way of war is appropriate only for superpowers, the United States unintentionally exported it to its smaller allies as RMA genetic material. RMA, one of the first doctrines that America did not borrow from others but developed on its own, necessarily assumes that one has a surplus of great power that lends it the capability to wage campaigns on its own terms, without the enemy having a vote or being able to act effectively at the same time against the strategic rear.

Against the West's techno-tactical and tactical might, Asia and the Middle East have an ancient and intensifying tradition of winning wars thanks to a strategy suited to the context and as a function of "the situation as a whole." And, indeed, Sun Tzu, Mao Tse-tung, Ho Chi Minh, Anwar Sadat, and Hassan Nasrallah succeeded in war even without attaining a military decision against their opponents, without bringing about the collapse of the original structure of the enemy force, and sometimes without participating in a classical manner in the prime major battle. In some cases, victory was achieved by simply not losing, and even not losing at only one level of the war.

Not surprisingly, the United States, Israel, and other Western nations usually demonstrate military superiority up to the operational level but do not realize their political objectives, and in that sense they lose the war (Algeria, Vietnam, the Yom Kippur War, the 2003 Iraq War, and the Second Lebanon War, to list just a few examples). Because of that particular blind spot, of not understanding the situation as a whole, Israel has not investigated its failure in the Second Lebanon War correctly. It has failed, among other reasons, because it viewed the war as not much more than a list of targets to attack with standoff fire, and therefore the main lesson it has learned is bringing the maneuver back to center stage of military activity. The maneuver and the direct tactical capabilities of force-on-force remain as crucial as they have always been, but if we merely seek to attack that list of targets using tanks and low trajectory firepower rather than by standoff fire, we have not learned much from the war. While restoring maneuver is crucial, and maneuver is an irreplaceable tool in the military toolbox and in shaping the nature of the war, the opposite of the notion of war as a list of targets to attack with standoff fire is not just maneuver; the opposite is also viewing war from a much broader and deeper perspective. The problem with a targeting list

# CONCLUSION

for standoff fire is mainly the narrow view of matters, and not just the lack of effectiveness of the fire in the circumstances of that particular war.

How do Israel's unique circumstances affect its particular national way of war? The traditional view hinges on adopting a defensive strategy that seeks to prevent, not generate, a change through war, along with an offensive mode of action meant to take the fighting into enemy territory, shorten the duration of the war, and test the sides' military effectiveness (strike force), not their stamina. This correct approach rests on three pillars – deterrence, early warning, and decision – and recently, there has been talk of adding a fourth pillar, namely, home front defense.

The deterrence pillar is problematic, as it is difficult to manage. Deterrence is based on conveying our capability of exacting a heavy war toll from the enemy, but in fact it takes place primarily in the enemy's mind. Clearly, should the enemy decide that it is willing to pay the anticipated price of the war, the weight of the deterrence factor will be diminished. This is precisely what Sadat decided on the eve of the Yom Kippur War. Deterrence also suffers from two additional difficulties: first, it is harder to deter third world dictatorships that are at times prepared to pay a steep price in economic terms and in human life, and even more difficult to deter non-state organizations; second, the validity of deterrence depends also on the enemy's war paradigm, and in certain war paradigms the relevance of our power (in absolute terms, or in relation to the enemy's power) is diminished, and consequently that of deterrence as well. Thus, for example, most American capabilities were irrelevant to the circumstances of the Vietnam War, and the strongest nuclear superpower did not manage to deter the Vietcong or North Vietnam. In order to maintain deterrence, we must define our power and our concept of force application in a way that will make it clear that we know how to exact a relevant, intolerable price from the enemy, and that we have found a response to the enemy's war paradigm.

This monograph does not deal in depth with issues related to early warning, but suffice it to say that early warning capability also depends on understanding the enemy's war paradigm. As in the 1973 war against Egypt, Military Intelligence in Israel failed to issue early warnings because it did not understand the Egyptian war paradigm, and the so-called "conception" relied on realizing a paradigm the Egyptians had not planned for and that in fact was never played out. Moreover, certain war paradigms can be realized even without intensive preparations, and therefore it is difficult to issue warnings about them.

The question of decision is the most complex, and is the cornerstone of this monograph. Decision is relevant primarily to short wars testing the sides' military effectiveness; decision is less relevant in a prolonged war where additional factors such as resources, stamina, and ability to enlist the international community are dominant. Israel would seem to have a relative advantage in military effectiveness, and usually a relative disadvantage in the other dimensions. The most suitable and correct war for Israel is a blitz in which maximum force is applied in order to realize the desired end state in a minimum of time, in order to deprive the enemy of its military capabilities or its strategic freedom of action, even before issues of stamina and resources surface, and before the international community becomes involved. Therefore, in order to attain decision, Israel's defense concept must focus on shaping and creating the circumstances that will cause a short war hinging on military effectiveness, and not get dragged into other types of war (such as happened wrongly in 2006 and to a degree in 2008). This process is dynamic and complex, because Israel's enemies understand this point very well and act to shape wars that deny it the opportunity for decision, while other factors, particularly stamina, become dominant. They attempt not to lose until they can outlast Israel. Gaining an advantage in this struggle, between Israel's desire to shape short wars of decision and the opponent's desire to shape prolonged wars of attrition, must stand at the heart of Israel's national defense concept.

Many of the aspects shaping the nature of a war have been surveyed above, including the ground maneuver. The maneuver is not simply an old fashioned techno-tactical way of getting ground assets in place inside their limited firepower envelope, but a basis for shaping the nature of the war. Maneuver can accelerate the war, hasten the clash between the forces and bring the test of their military effectiveness to a climax, and rapidly deprive the enemy's strategic freedom of action. Maneuver towards a strategic center of gravity is liable to compel the enemy force to abandon its paradigm of exhaustion and guerilla, and assume positions for a classical major defensive battle that it had sought to avoid. Without maneuver, a prolonged attack with reciprocal fire may develop, and in effect become a war of attrition testing the sides' stamina or inviting international intervention. At the same time, the maneuver must be designed correctly. A wrong maneuver – for example, a frontal assault on a saturated defensive deployment – is liable to lead to attrition and stalemate and test the sides' stamina (as happened in the trenches of World War I). Even a tactical maneuver may end with operational and strategic-level exhaustion, and therefore it is necessary to understand the effects of all levels of the war and its axes on the design of the maneuver. The maneuver must create the desired advantage and undermine the enemy's war paradigm – tactically, operationally, and strategically.

The addition of a home front defense pillar to Israel's defense concept is also not a matter of course. As argued in this study, the advantage of rockets and missiles is not in their direct physical effectiveness, rather in the strategic effect caused by the disruption to normal civilian life over time and the undermining of the cohesion between the government, the public, and the military. Such an outcome is not directly dependent on the number of rockets that fall; as long as the enemy retains residual defiance capability, it might very well attain its objectives. However, no interception system that exists today is capable of providing a hermetic defense, and it is enough that a relatively limited number of rockets penetrate for the enemy to continue to maintain residual defiance capability, while the investment in an interception system commands resources that would be better spent in the primary effort to achieve decision. There is also asymmetry between the relatively modest investment required to increase the rocket threat and its penetration capability, and the enormous investment needed to improve the interception response. Apparently an arms race between the rocket and the interceptor will almost always favor the rocket. At least in the techno-tactical world of today, it seems that the response to the rocket threat is not interception, but rather reducing the duration of the home front's exposure to fire by achieving a quick decision against the enemy. The more precise the rockets and missiles and the more the nature of the ensuing threat worsens, the greater the need for a quick and violent decision.

Some additional crucial points regarding Israel's unique defense circumstances should be mentioned. Israel must refuse (insofar as it is practical) to wage wars of gradual escalation, and it must not allow the enemy to determine the outline, duration, location, or intensity of the war. Israel should not react proportionately to the enemy's actions, because by doing so it enables the enemy to dictate the nature of the war whereby Israel is liable to be dragged into a prolonged exchange of blows, i.e. attrition. Exchanges of fire, raids, and a long list of limited operations channel the confrontation towards tests in which Israel has no interest. Creating attrition requires time, and hence shortening the duration of the war must be a key factor in Israel's national defense concept. Therefore, from the moment an armed conflict breaks out, Israel must take control of the conflict's outline, redefine its geography and its intensity, and apply a maximum of force in a minimum amount of time in order to attain a decision, or at least to deny the enemy its strategic freedom to continue fighting.

## The Right Lessons from Lebanon 2006

The tension between realizing Israel's decision effort and disrupting the enemy's attrition effort also affects the allocation of resources. It is preferable to build up a strike force and take action to realize the Israeli decision effort rather than build up defensive capabilities or allocate forces to disrupt the enemy's opposite parallel campaign. Foiling the enemy's strategy is accomplished by quick strategic decision, and not through a direct

Sisyphean effort against systems such as surface-to-surface missiles, surface-to-surface rockets, or guerilla warfare. Rapid decision requires boldness, risk-taking, subterfuge, surprise, and deception. It is also important to preserve a reservoir of versatile and varied military capabilities: when the mode of applying force is one-dimensional – such as the "all torpedo ships" approach of the Jeune École, the "all tanks" approach on the eve of the Yom Kippur War, or the "all standoff fire" approach on the eve of the Second Lebanon War – it is easier for the enemy to adjust and develop a war paradigm that will deconstruct and neutralize the effectiveness of our capabilities. Military effectiveness and decision constituents are always context-dependent, and must be suited to the particular war paradigm.

The experience of the summer of 2006 might also prompt regular state enemies to adopt the Hizbollah war paradigm. Such a development would present Israel with new challenges, and in particular, with the difficulty of defining what constitutes decision, the center of gravity to be attacked, and how nevertheless it imposes a prime major battle instead of being caught up in a war of attrition lacking a finish line. It is difficult to impair the enemy's capability of operating effectively when the enemy relies on disappearing, low signature, high redundancy formations that achieve their objectives as long as they maintain their residual defiance capability. As war becomes more asymmetrical, so the enemy's concentrated military masses are liable to disappear, or at least lower their signature. Whoever applies military force using a paradigm that does not reflect this reality is liable at best to be striking a punching bag filled with feathers and at worst, striking at a punching bag filled with nails. In either case, the punching bag will not crush. In such a case, it is necessary to identify alternate, more complex centers of gravity. These are likely to be the enemy's strategic and operational basic assumptions that when undermined destabilize the enemy's war paradigm. In addition, they are likely to be political centers of gravity, which when struck undermine the enemy's governmental stability or its legitimacy. The growing complexity of war compels us to deal with defeating the enemy's war plans, no less than – and possibly more than – attacking the enemy's mass and assets. There is no greater achievement for a military leader than the enemy suddenly discovering that the war actually being waged is completely different from the one the enemy planned for. The difficulty in defining decision and centers of gravity does not mean that these terms should be abandoned, only that they are complex and must be approached as such. As analytical tools, they are still more crucial than ever to the process of planning war.

Israel suffers from a relative disadvantage in its ability to leverage the international community to determine the outcome of the war, but this disadvantage is not absolute. Particularly regarding enemies such as

# CONCLUSION

Hizbollah, Hamas, Syria, and Iran, Israel in the current political reality might be able to enjoy a diplomatic tailwind. In the past, the political dialogue would begin with the end of the fighting, and the political end state would to a large extent reflect the military reality at the end of the battles. The connection between the success in terms of military outcome and the political achievement afterwards was very close. Today that connection is loosening, and just like every other level of the war, the political level too has its own autonomous rationale. This reality represents a problem but at the same time an opportunity. Some of the military clashes are foreseeable ahead of time, sometimes years in advance, and it is possible to try to maximize the political achievement even before the outbreak of war, and reach discreet understandings with central players in the international arena. Security Council Resolution 1701 (and the difficulties of its implementation) did not stem from the outcome of the battles, but primarily reflected the complex balance of political power between Israel, the United States, France, the Sunni Arab states, the Sunni and Christian factions in Lebanon, Hizbollah, the Lebanese Shiite community, Syria, and Iran. However, Israel began its diplomatic campaign only when the battles began. For years, Israel avoided initiating a dialogue, even with its friends – the United States and the European nations – designed to coordinate in advance the nature of the arrangement to be reached after the anticipated and unavoidable conflict with Hizbollah. Had Israel begun a diplomatic campaign before the military campaign began, it might have been possible to attain a better political outcome, and one unlinked to the military outcome.

## The Last Word

The complexity of the war compels us to contend with the broader picture – "the situation as a whole." We must understand the war at all its levels – the long term national and political objectives, the grand strategy, the military strategy, the systemic-operational planning, the tactical battle, the techno-tactics, the logistics, the information, the narrative, and, of course, the buildup of force and the concept of how to apply it. In order to win a war that is ever more complex and losing clear boundaries, we must demonstrate superiority at every level and in every aspect. Perhaps most of all, we must understand a war in its distinctive context. We must understand what issues the enemy, for its own reasons, wants to test in a particular impending war, what "self-evident" basic assumptions the enemy's plans rest on, how we can undermine the enemy's basic assumptions, and how we can force other issues more suited to us to be put to the test. In this respect, we are still in our infancy.

"Whoever brings the danger to his opponent demonstrates better fortitude than whoever repels the danger. What is more, the fear of the unknown grows as a result. When you penetrate the enemy's land, you can clearly see the enemy's strengths and weaknesses."

Scipio Africanus

# Notes

1. In the context of this monograph, the term "attrition" has different meanings in different contexts. Attrition here is more consistent with the familiar definition of a wearing down or weakening of resistance, especially as a result of continuous pressure or harassment, rather than being synonymous with statistical "non-smart" destruction.
2. In this memorandum, ground mobile warfare is defined as maneuver, although there are other accepted definitions for "maneuver."
3. This description reflects the popular perception of Clausewitzian theory, although from his extensive body of writing a more complex picture emerges.
4. See also Ron Tira, "Does Israel Win Its Wars?" *Ma'arachot* 407, IDF Publishing House (June 2006): 4–9.
5. RMA is an American concept that argues that the nature of war has undergone an essential change as a result of the convergence of information technologies and communications, sensory technologies, and precision strike technologies, along with a dramatic change in military doctrine and the systemic and organizational approaches to warfare. For more, see Chapter 1.
6. Carl von Clausewitz, *On War* (Princeton: Princeton University Press, 1984), p. 596.
7. Clausewitz, *On War*, p. 577.
8. Clausewitz, *On War*, p. 487.
9. Clausewitz, *On War*, p. 258.
10. FM 3-0, Department of the Army, February 2008; JP 3-0, Joint Chiefs of Staff, February 2008.
11. Avi Kober, *Military Decision in Israel-Arab Wars 1948–1982*, 4th ed. (Tel Aviv: Ma'arachot, 2001), pp. 25–26.
12. Kober, *Military Decision in Israel-Arab Wars*, p. 121.
13. Israel Tal, *National Security: The Few against the Many* (Tel Aviv: Dvir, 1996), p. 57.
14. Kober, *Military Decision in Israel-Arab Wars*, p. 172, quoting a statement by Maj. Gen. Menahem Einan.
15. Arden Bucholz, *Moltke and the German Wars 1864–1871* (New York: Palgrave, 2001), pp. 154–59.
16. Shimon Naveh, *The Creation of Military Excellence*, 3rd ed. (Tel Aviv: Ma'arachot, 2003), pp. 36–39.
17. Air Force Doctrine Document 2-1.2, Secretary of the Air Force, June 2007.
18. Air Force Doctrine Document 2-1.2, Secretary of the Air Force, September 2003, pp. 2–3.
19. Ibid., p. 7.
20. David Johnson, *Learning Large Lessons: The Evolving Roles of Ground Power and*

*Air Power in the Post-Cold War Era* (Arlington, VA: Rand Corporation, 2007), pp. 82–89.
21 Ron Tira, *The Limitations of Standoff Firepower-Based Operations: On Standoff Warfare, Maneuver, and Decision* (Tel Aviv: Institute for National Security Studies, Memorandum No. 89, 2007).
22 The concept of EBO seeks to locate vulnerability nodes in the enemy system and to attack them in a way that suppresses operational effectiveness and denies its rationale.
23 Clausewitz dealt with the ideas of the Revolution in Military Affairs (RMA) of his time, such as Dietrich Heinrich Freiherr von Bülow's thesis that in the transition from cold weapons to hot weapons, logistics became the element dictating the outlines of the maneuver. Bülow claimed that it is possible to analyze logistics using geometrical parameters and thus it is possible to develop an exact science of war, at whose center stands the geometry of supply lines and their disruption, without the need for conducting a major prime battle. For more on these topics, see Azar Gat, *Sources of Modern Military Thinking* (Tel Aviv: Ma'arachot, 2000).
24 Clausewitz, *On War*, p. 228.
25 For sources and more on the topic, see Samuel R. Williamson, *The Politics of Grand Strategy: Britain and France Prepare for War, 1904–1914* (London: Ashfield Press, 1990).
26 The Jeune École concept was another RMA of its time, and in its principles is reminiscent of Israeli revolutionary approaches, such as the "dynamic molecule" and "decentralized warfare," which like the Jeune École, also failed as a main method of battle though they certainly have a role to play as an additional element in the context of a more balanced force. The Jeune École concept posited that in the age of steam, the torpedo, and the submarine, large naval task forces with their central heavy ships no longer had any role to play. Rather, the age of "swarms" had arrived – the age of small, quick vessels armed with advanced weaponry. These small vessels would avoid large battles at sea, and would instead operate by means of "stings" or try to create indirect effects by interfering with maritime supply lines and disrupting the enemy's economy. The Jeune École view was implemented in the French navy in a revolutionary and immediate way, i.e., not through a gradual, evolutionary development, and as a result France lost its status as the second most important naval power in the world, instead becoming an inferior adversary in comparison to the other large European navies (though the Jeune École did have successes against colonial enemies, such as the Chinese navy). And indeed, France reached the conclusion that only the destruction of the enemy's navy in major battles – not "stings" or blows to naval trade – would ensure naval superiority and victory. France also understood that the guerilla warfare of the Jeune École was suited to weak countries adopting the strategy of defensive disappearance, and not to navies seeking maritime dominance. The French navy even recognized that the Jeune École viewed the world through the lenses of technology and technotactics, in a conflict in a vacuum between a rapid naval vessel and a heavy ship, without understanding the broader contexts of deploying large, varied, and balanced task forces in a campaign setting. The Jeune École concept also did

## NOTES

not take into consideration the response to the small torpedo ship, in particular the development of the destroyer. However, at the late stage at which France understood it had erred, the economic and political burden of rebuilding a balanced navy was enormous, and France found it difficult to implement its new-old policy.

27  Israel of course surprised the Egyptians by attacking first, while the Egyptians also relied on their stationary fortifications.
28  To be precise, at that point the designated purpose of the Syrian forces was the defense of Damascus, and were we to claim that Israel could have threatened Damascus, it would mean that the Syrian forces had lost their capability to fulfill their designated purpose, namely, decision was achieved against them from the defensive point of view. To claim that the Syrian forces continued to maintain their defensive capability is to say that at that point Israel was still not able to threaten Damascus.
29  This, although it is possible to claim that Israel never succeeded in leveraging a military decision into a victory (the realization of the political objective). See Tira, "Does Israel Win Its Wars."
30  For sources and further information about the higher levels of World War II, see Richard Overy, *Why the Allies Won* (New York: Norton, 1997).
31  The bombing of German industries also forced the Germans to redirect the Luftwaffe to missions of defending the German skies, thus abandoning the fronts.
32  For example, in 1939 Germany produced some 1,300 tanks, and the US, Britain, and the Soviet Union together produced 3,900, while in 1943, Germany produced 17,300 tanks, and the US, Britain, and the Soviet Union produced some 61,100 tanks. In 1939, Germany and Japan together produced 12,800 aircraft, and the US, Britain, and the Soviet Union produced 24,200, but in 1944, Germany and Japan produced 68,000 aircraft, while the US, Britain, and the Soviet Union produced 164,000.
33  Important background facts include also the preference of Western military commanders to invest resources in logistics and military organization, an investment out of all proportion to the corresponding investment of the Axis nations. In the Pacific campaign, the Americans had 18 support, logistics, and administration personnel for every fighter; in the Japanese military, the ratio was one to one. In 1944, the logistics of the American and British armies were fully mechanized, whereas Germany was still using some 1.25 million horses. The Allies' pace of training new fighters was also significantly faster than that of the Axis, as was the capacity to learn: the Germans and the Japanese ended the war with almost the same doctrine they espoused at its beginning, whereas the US, Britain, and the Soviet Union learned, adapted, and changed their doctrine dramatically.
34  For sources and further discussion, see Danny Asher, *Breaking the Conception* (Tel Aviv: Ma'arachot, 2003); and Saad Eldin el-Shazly, *The Crossing of the Canal* (Tel Aviv: Ma'arachot, 1987).
35  Egypt's estimations regarding its anticipated losses in crossing the canal and in the ensuing battles were in fact significantly higher than the actual losses. In addition, the Yom Kippur War broke out only three years after the end of the

War of Attrition, in which Israel exacted a steep price from the Egyptian home front, with attacks on national infrastructures like oil refineries, and harsh blows against cities along the canal, turning millions of Egyptians into refugees.

36  A presidential directive to the Commander-in-Chief of the Egyptian Armed Forces, October 1, 1973.
37  Strategically – because Military Intelligence relied on a conception whereby Egypt would not embark on a war as long as it could not create aerial superiority in the depth of the Sinai, but the shift in the Egyptian paradigm to a limited bite-and-hold under the cover of the Integrated Air Defense System of the western bank of the canal rendered that conception irrelevant. Operationally and tactically – because crossing the canal by any Egyptian unit in a sector in which it was defensively deployed and operating out of defensive positions, without significant deployment movements, greatly reduced the time needed to prepare an attack as well as the extent of the preliminary activities.
38  This was also an expression of a dispute within the IDF itself regarding the designated function of the outposts: were they a defensive line or merely an early warning line.
39  Shazly, *The Crossing of the Canal*, p. 194.
40  Egypt's successful operational concept also challenges the principle of depth in "operational art." The Egyptian crossing of the canal was linear and purposely lacking a significant component of depth, and it was the lack of Egyptian depth on the eastern bank that led to the defeat of Israel's operational thinking and even to the Egyptian success at the strategy and grand strategy levels. It would also seem as if the only way the IDF could have defeated the Egyptian paradigm would have been to deploy the main mass of its forces along the Suez Canal in a dense linear formation with no depth, with a "simple" destruction of the mass of crossing forces.
41  See n. 36; and Shazly, *The Crossing of the Canal*, pp. 116, 130–31.
42  For further discussion, see Tira, *The Limitations of Standoff Firepower-Based Operations*.
43  For sources and further discussion see also Haggai Golan and Shaul Shay (eds.), *As the Engines Rumbled* (Tel Aviv: Ma'arachot, 2006).
44  The decision against the MCP (the Chinese-Communist guerilla organization) and the orderly transfer of all of Malaya (including the western provinces where the Chinese were a majority) into the hands of the pro-British Muslim Malayans.
45  Ian E. W. Beckett, *Modern Insurgencies and Counter-Insurgencies* (London: Routledge, 2005), pp. 190–91.
46  Harry G. Summers, *On Strategy: A Critical Analysis of the Vietnam War* (Novato, California: Presidio Press, 1982), p. 1.
47  As far as they relate to the uprooting of the PLO from Lebanon. Of course the more far reaching objectives of establishing a friendly Christian regime in Lebanon and the political disappearance of the PLO were not attained.
48  It is not easy to find examples that conform to Sun Tzu's approach, but perhaps President Reagan's SDI project is an example of attack on the enemy's strategy that ended with success without the actual use of force. The basis of

the strategic relations between the superpowers during the Cold War was the assumption that both sides were able to absorb a first nuclear strike and nonetheless maintain enough residual capability to inflict a second blow that would completely annihilate the enemy. In this manner, nuclear war guarantees Mutually Assured Destruction (MAD), and therefore neither side would begin a nuclear war. The Strategic Defense Initiative (SDI, more commonly known as Star Wars) was meant to provide the United States with the ability to intercept nuclear missiles still in flight, and thus stop a Soviet first strike, or – even more than that – deny the Soviets second strike capability, and thus undermine the basic assumption of MAD. Therefore the SDI challenged the Soviet Union with two intolerable alternatives: end the era of the nuclear balance of deterrence with the Soviet Union at a disadvantage, guaranteeing its strategic inferiority, or enter an incalculably expensive arms race that could very well push the Soviet Union into bankruptcy. As with every clever strategic move, any Soviet response to this dilemma was good for the Americans. The SDI thus attacked the strategic assumptions of the Soviet Union and the capabilities of its national resources, and there are those who claim that this contributed towards the collapse of the Soviet Union.

49  Mao Tse-tung, *Selected Military Writings* (Peking: Foreign Language Press, 1963), pp. 81–82.
50  B. H. Liddell Hart, *The Strategy of Indirect Approach* (Tel Aviv: Ma'arachot, 1956), Chapter 14.
51  Liddell Hart, *The Strategy of Indirect Approach*, p. 334.
52  Liddell Hart's analysis focuses on Britain's added value as a member of a coalition of fighting nations. While not mentioned explicitly, it seems that Liddell Hart does not claim that it was possible to win World War I by naval blockade alone and without some forces taking position in the battlefield on the ground against the German forces.
53  A successful application of the strategy of indirect approach in the West is General William Sherman's March to the Sea (1864) during the American Civil War. The Civil War was one of the few wars waged between two democracies, and as such, public support for the war was critical to both sides. The March to the Sea was a campaign by the North, intended to undermine Southern public support for the war, and in particular in the South's political center of gravity – Georgia and the two Carolinas. General Sherman, who commanded the March, sought to bring the war home to the civilians at the heart of the South's political power and thereby destroy the sources feeding the war effort. The method of action combined pinning of the South's forces to the front during prolonged attacks while maneuvering deep behind Southern forces over 450 km of the South's strategic rear; Sherman left behind him nothing but scorched earth, including the burning of Atlanta and Savannah. In fact, the March to the Sea made an essential contribution to the collapse of the Southern will to fight and the collapse of General Lee's army.
54  An interesting historical anecdote: The Prussian/German war ethos presented the Battle at Cannae as the ideal model for imitation. In this tremendous clash between two masses during the Second Punic War, Hannibal did achieve an impressive tactical victory over the Romans. Nonetheless, when all was said

and done, the Roman superiority at the strategic and operational levels caused Hannibal to lose the Second Punic War, a fact disregarded by the Prussian/German ethos. This hints at the prism of tactical thinking that in many cases is characteristic of Western militaries.

55 For the sake of historical accuracy: the threat created by Scipio against Carthage forced Hannibal to abandon Italy and set sail for his capital to defend it, so that in the end, the two forces did meet in the same theater of operations.
56 For further discussion, see Isaac Ben Israel, *The First Missile War – Israel-Hizbollah* (Tel Aviv: Tel Aviv University, 2007).
57 A large portion of the rocket force was disposable, so that in any case, there was little point in responding to the sources of the rocket fire with fire. These launchers were also operated by timers, so that there wasn't even an operator present who could be attacked.
58 These are methods of warfare in which small forces with a low signature operate or direct precision strikes from a distance. Such methods certainly have their place as a proportional piece within the construction of a balanced force and within a menu of operating varied forces.
59 Giora Romm, "A Short History of the Second Lebanon War," in Essay Collection no. 32, The Fisher Institute, 2006.
60 The problem of a direct, massive attack on the civilian home front is at this point unique to Israel, but potential enemies of the West are capable of token attacks on Europe, the United States, and Russia, and are capable of massive attacks against South Korea, Japan, American assets in the Persian Gulf, and against the moderate Gulf states. As the range and precision of ballistic weapons increase, this problem may represent an even more acute threat against the West.
61 Benjamin S. Lambeth, *NATO's Air War for Kosovo: A Strategic and Operational Assessment* (California: Rand Corporation, 2001).
62 Johnson, *Learning Large Lessons*, pp. 114–15.
63 Tira, *The Limitations of Standoff Firepower-Based Operations*, pp. 29–32.
64 Michael I. Handel, *Masters of War – Classical Strategic Thought*, 3rd ed. (London: Frank Cass, 2001), p. xxii.
65 Thanks to Brigadier General (ret.) Oded Tira for the simile.
66 Paul Kennedy, *Grand Strategies in War and Peace* (New Haven: Yale University Press, 1991), p. 3.

# Index

abstract centers of gravity
  defined, 66, 115
  guerilla paradigm, 115
  Morocco crisis (1904–6), 28
  Second Lebanon War (2006), 92, 96, 97
  US doctrine, 16
  World War II, 42, 116
  Yom Kippur War (1973), 55, 57, 116
  see also centers of gravity; military centers of gravity; physical centers of gravity
Aden, 70
Afghanistan, 70
Algeria, 61, 70, 71, 87
Arab states
  grand strategy, 19, 43–4
  military doctrines, 23, 42, 78–9
Arab–Israeli wars see Sinai Campaign; Six Day War; War of Attrition; Yom Kippur War; First Lebanon War; Second Lebanon War; Operation Cast Lead
Asia, military doctrines, 4, 78–9, 125
asymmetrical wars, 3–4, 37–8, 65–7
  centers of gravity, 37, 38, 66–7
  civilian–political stamina, 65, 66, 67
  classical Clausewitz doctrine, 37
  government–military–civilian triangle, 66
  grand strategy level, 65, 66
  guerilla paradigm, 111, 114–15
  international community, 65
  military decision, 37, 65, 66
  military effectiveness, 37, 65, 66
  military end state, 65
  military strategy level, 65, 66
  non-state opponents, 69–76
  operational centers of gravity, 66
  operational level, 65, 66
  political end state, 3, 65
  resources, 65
  Second Lebanon War (2006), 85–102
  Sinai Campaign (1956), 61–5
  theater of operations, 4, 37, 38, 65
  War of Attrition (1969–70), 59–60
  World War II, 3, 38–42
  Yom Kippur War (1973), 33, 42–58
attrition, 1
  attack on assets, 7, 67
  British doctrine, 124
  defined, 132$n$
  guerilla paradigm, 112, 115
  Israeli doctrine, 5–6, 22, 127, 128, 129
  Operation Cast Lead (2008–9), 106
  Second Lebanon War (2006), 93, 94
  US doctrine, 5, 21, 22
  tactical maneuver, 119, 127

Bin Laden, Osama, 92
Bismarck, Otto von, 27
blitzkrieg, 38, 39, 47, 54
Boer War, 70
Britain
  Malaya campaign, 70, 71, 87, 135$n$
  military doctrine, 79–80, 124
  Morocco crisis (1904–6), 27–9
  Napoleonic Wars, 80
  non-state opponents, 70
  Sinai Campaign (1956), 61–5
  World War I, 79–80
  World War II, 38–40, 57, 117, 124
Bülow, Dietrich Heinrich Freiherr von, 133$n$

Carthage see Punic War, Second
centers of gravity
  analytical tool, as, 7, 16, 129
  asymmetrical wars, 37, 38, 66–7
  classical Clausewitz doctrine, 1, 7, 11, 15–16, 25–7
  defined, 7, 17
  Dien Bien Phu campaign (1953–54), 74
  First Lebanon War (1982), 75, 76
  guerilla paradigm, 8–9, 113–14, 115–21, 122
  Morocco crisis (1904–6), 28

# INDEX

non-state opponents, 71
Operation Cast Lead (2008–9), 102, 104, 107
operational art, 25–6
parallel warfare, 3, 82
Revolution in Military Affairs (RMA), 20, 21, 25–6
Second Lebanon War (2006), 76, 86–90, 91–2, 94–5, 96–7, 100
Second Punic War, 82, 118
Sinai Campaign (1956), 63–4
Six Day War (1967), 31, 32–3, 116
Systemic Operational Design (SOD), 19
US doctrine, 16–17, 20–1
War of Attrition (1969–70), 59–60
World War II, 42, 116, 117, 118
Yom Kippur War (1973), 33, 34, 35, 55–7, 116, 118
*see also* abstract centers of gravity; military centers of gravity; physical centers of gravity
Chechnya, 70, 87
Chinese civil war, 70, 78
Churchill, Winston, 40
civilian–political stamina, 4–6
  Arab–Israeli wars, 42
  asymmetrical wars, 65, 66, 67
  France, 12
  guerilla paradigm, 112, 114, 115, 119, 120
  Iraq wars, 24
  Israeli doctrine, 5–6, 18, 19, 21–2, 24, 126, 127
  maneuver (ground mobile warfare), 127
  Kosovo War, 24
  Morocco crisis (1904–6), 28
  non-state opponents, 72
  outcome of the war determination, 2
  Prussian/German doctrine, 18, 123
  Second Lebanon War (2006), 8, 93, 94–5
  Second Punic War, 82
  Six Day War (1967), 32
  US doctrine, 5, 21–2
  Vietnam War, 2, 73, 99–100
  War of Attrition (1969–70), 59–60
war matrix, 9
wars of attrition, 5
World War II, 2, 12, 38, 40, 42

Yom Kippur War (1973), 2, 45, 52, 56, 57, 100
Clark, Wesley, 23
classical Clausewitz doctrine, 2–3, 11–13
  asymmetrical wars, 37
  centers of gravity, 1, 7, 11, 15–16, 25–7
  First Lebanon War (1982), 76
  military decision, 1, 6, 11, 16, 25–7
  military end state, 51, 65
  non-state opponents, 74
  political end state, 3, 65
  political objectives, 1, 11, 15
  Six Day War (1967), 31, 33, 35
  theater of operations, 15–16
  Yom Kippur War (1973), 33, 34, 35, 53
Clausewitz, Carl von, 31
  annihilation principle, 15, 25, 55, 120
  guerilla warfare, 70
  Revolution in Military Affairs (RMA), 133$n$
  wondrous trinity, 44, 86
  *see also* classical Clausewitz doctrine
Cronkite, Walter, 73
Cyprus, 70

Dahiya quarter, Beirut, 91
Damascus, 33, 34, 134$n$
decisive points, US doctrine, 16–17
dictatorships, 8, 57, 67, 93, 126
Dien Bien Phu campaign (1953–54), 4, 74
Douhet, Giulio, 20
Dutch East Indies, 70

Effect Based Operations (EBO), 25–6, 56, 113, 117, 133$n$
Egypt
  Operation Cast Lead (2008–9), 103, 104
  Sinai Campaign (1956), 61–5
  Six Day War (1967), 32–3, 42, 116
  War of Attrition (1969–70), 59–60, 134–5$n$
  weapons smuggling into Gaza, 103, 104
  Yom Kippur War (1973), 2, 3, 33, 42–58, 112, 116, 126, 134$n$, 135$n$
Eisenhower, Dwight D., 62, 124
end states *see* military end state; political end state

139

# INDEX

Fabius Maximus, 80
Fatah, 104
fighting capability, 25, 66, 67
  American doctrine, 16
  defined, 17
  Egyptian forces, 57
  Hizbollah, 91
  Israeli forces, 64
  Syrian forces, 34
  Vietcong, 72
France
  Algerian campaign, 61, 70, 71, 87
  civilian–political stamina, 12
  Indochina campaign, 70, 74
  Jeune École concept, 133–4$n$
  Morocco crisis (1904–6), 27–9
  Napoleonic Wars, 51, 70, 80
  Sinai Campaign (1956), 61–5
  World War II, 3, 38–9, 99, 116, 117

Gamasi, Abdul Ghani, 43
Gaza
  Israeli withdrawal, 12
  Operation Cast Lead (2008–9), 1, 102–8
Germany
  civilian–political stamina, 18, 123
  military decision, 18
  military doctrine, 18–19, 123, 136–7$n$
  Morocco crisis (1904–6), 27–9
  multi-front wars, 18–19
  resources, 18
  World War I, 79–80
  World War II, 2, 3, 26, 38–40, 41–2, 54, 57, 70, 99, 116, 117, 119, 134$n$
Gibraltar, 27, 28
government–military–civilian triangle
  asymmetrical wars, 66
  Clausewitz's wondrous trinity, 44, 86
  guerilla paradigm, 112
  Israel, 44, 52, 56, 86–90, 94, 101
  Vietnam War, 73
grand strategy level, 2–3, 9
  Arab states, 19, 43–4
  asymmetrical wars, 65, 66
  Israel, 19
  Morocco crisis (1904–6), 27
  Operation Cast Lead (2008–9), 104
  Second Lebanon War (2006), 100
  Sinai Campaign (1956), 61–2, 65
  Six Day War (1967), 33
  Vietnam War, 73, 74

World War II, 39–40, 41–2
Yom Kippur War (1973), 3, 43–4, 48, 50, 51, 52, 54–5
Greece, 70
Guderian, Heinz, 124
guerilla paradigm, 8–9, 109–22
  asymmetrical wars, 111, 114–15
  attrition, 112, 115
  centers of gravity, 8–9, 113–14, 115–21, 122
  civilian–political stamina, 112, 114, 115, 119, 120
  Clausewitz and, 70
  First Lebanon War (1982), 75–6
  Hizbollah, 1, 75–6, 87–90, 96, 107–8, 114
  maneuver (ground mobile warfare), 116–19
  military decision, 112–13, 114–15, 120–2
  military strategy level, 110–15
  operational level, 110–15
  parallel warfare, 114–15
  residual defiance capability, 6, 76, 114, 115, 128, 129
  Revolution in Military Affairs (RMA), 110–13
  "winning by not losing" principle, 110, 115
  *see also* Palestine Liberation Organisation (PLO); Viet Minh; Vietcong
Gulf Wars *see* Iraq

Hamas, Operation Cast Lead (2008–9), 102–4, 105, 106–8
Handel, Michael, 121
Hannibal, 4, 7, 80, 81–2, 116, 136–7$n$
Hizbollah
  guerilla paradigm, 1, 75–6, 87–90, 96, 107–8, 114
  residual defiance capability, 6, 90, 94, 114
  Second Lebanon War (2006), 6, 8, 75–6, 85–102, 129
Ho Chi Minh, 78, 124, 125
Hussein, Saddam, 92

Indochina, 70, 74
international community, 1–2, 4, 11
  Arab–Israeli wars, 42
  asymmetrical wars, 65

# INDEX

guerilla paradigm, 115–16
Iraq wars, 24
Israeli approach, 5, 119, 127, 129–30
Kosovo War, 24
Operation Cast Lead (2008–9), 104–5, 106
Second Lebanon War (2006), 94, 100, 130
Sinai Campaign (1956), 61, 62
war matrix, 9
wars of attrition, 5
Yom Kippur War (1973), 2, 42, 45, 51, 100
Iran, 62
  Second Lebanon War (2006), 96
Iraq
  War (1991), 18, 23, 111, 116
  War (2003), 23, 70, 92, 111, 116, 118
Israel
  civilian–political stamina, 5–6, 18, 19, 21–2, 24, 126, 127
  defense dialogue with US, 23
  deterrence pillar, 19, 43–4, 52, 55, 126
  geo-strategic position, 5, 22, 87
  government–military–civilian triangle, 44, 52, 56, 86–90, 94, 101
  home front defense pillar, 101–2, 127–8
  international community, 5, 119, 127, 129–30
  maneuver (ground mobile warfare), 125
  military decision, 17–18, 19, 126–7
  military doctrine, 5–6, 17–18, 19, 22–3, 24–5, 125–8
  military effectiveness, 5–6, 126–7
  multi-front wars, 18
  resources, 5, 18, 23, 24
  Sinai Campaign (1956), 61–5
  Six Day War (1967), 31–3, 42, 99, 116
  standoff based wars, 5, 6, 8, 24, 88, 90, 93, 101, 125–6
  techno-tactical superiority, 4
  undesirable wars of attrition, 5–6, 22, 127, 128, 129
  War of Attrition (1969–70), 59–60, 134–5n
  withdrawal from Gaza, 12
  see also Operation Cast Lead; First Lebanon War; Second Lebanon War; Yom Kippur War
Israel Defense Forces (IDF)
  First Lebanon War (1982), 75, 76

Operation Cast Lead (2008–9), 102, 105, 107
precision arms technology, 121
Second Lebanon War (2006), 8, 76, 88, 90–4, 95–6
Sinai Campaign (1956), 65
Six Day War (1967), 32
War of Attrition (1969–70), 59
Yom Kippur War (1973), 33–4, 43, 45, 46–8, 50, 52, 53, 54–5, 56, 112, 135n
Italy, World War II, 40

Japan, World War II, 40–1
Jeune École, 28, 129, 133–4n
Johnson, Lyndon, 73
Jordan, expulsion of PLO, 75

Kennedy, Paul, 124
Kenya, 70
Kissinger, Henry, 51
Kober, Avi, 17, 25
Kosovo, 23–4, 58, 117–18
Kuwait, Gulf War (1991), 18

Lebanon, Shiite community, 97
Lebanon War, First (1982), 4, 75–6, 87–8, 95, 96
Lebanon War, Second (2006)
  attrition, 93, 94
  centers of gravity, 76, 86–90, 91–2, 94–5, 96–7, 100
  civilian–political stamina, 8, 93, 94–5
  Hizbollah's paradigm, 6, 8, 75–6, 85–102, 129
  international community, 94, 100, 130
  Israel Defense Forces (IDF), 8, 76, 88, 90–4, 95–6
  Israeli approach, 1, 8, 12, 25, 85–102, 120, 125, 128–30
  Israeli legitimacy, 12
  military effectiveness, 93, 94
  military end state, 93, 94, 98, 99
  military strategy level, 89, 90, 91, 92
  operational level, 88, 89, 90, 92, 95
  political end state, 3, 92, 97–8, 99
  systemic-operational design level, 90–1, 97, 98, 118, 130
  tactical level, 90, 95, 101
  techno-tactical level, 90, 92, 101
  "winning by not losing" principle, 5, 94
Liddell Hart, B.H., 79, 80, 124, 136n

# INDEX

logistical level
   Sinai Campaign (1956), 64–5
   World War II, 39, 41
logistics, 133$n$, 134$n$

Maginot Line, 38, 39, 42, 54
Malaya, 70, 71, 87, 135$n$
maneuver (ground mobile warfare), 8, 127
   defined, 132$n$
   guerilla paradigm, 116–19
   Israeli approach, 125
   Operation Cast Lead (2008–9), 104, 105–6
   Second Lebanon War (2006), 95
Mao Tse-tung, 4, 78, 125
Marshall, George, 124
Middle Eastern cultures, military doctrines, 4, 125
Midway, Battle of, 40, 42
military centers of gravity
   classical Clausewitz doctrine, 1, 11
   Yom Kippur War (1973), 55–6
   *see also* abstract centers of gravity; centers of gravity; physical centers of gravity
military decision
   analytical tool, as, 6–7, 129
   asymmetrical wars, 37, 65, 66
   classical Clausewitz doctrine, 1, 6, 11, 16, 25–7
   defined, 1, 17, 83
   defining of elements, 100
   Dien Bien Phu campaign (1953–54), 74
   First Lebanon War (1982), 75, 76
   German/Prussian approach, 18
   guerilla paradigm, 112–13, 114–15, 120–2
   Israeli approach, 17–18, 19, 126–7
   military effectiveness, 2, 126–7
   operational art, 25–6
   Revolution in Military Affairs (RMA), 25–6
   Second Lebanon War (2006), 94
   Second Punic War, 81
   superpowers, 18
   US approach, 17, 18
   Vietnam War, 73–4
   wars of attrition, 5
   World War II, 39, 42

Yom Kippur War (1973), 34–5, 50, 51, 52, 53–5, 112
military effectiveness
   asymmetrical wars, 37, 65, 66
   classical Clausewitz doctrine, 1
   dependence on prevailing paradigm, 7
   guerilla paradigm, 114
   Israeli preference for, 5–6, 126–7
   military decision, 2, 126–7
   Second Lebanon War (2006), 93, 94
   Second Punic War, 82, 99
   Sinai Campaign (1956), 61
   Six Day War (1967), 32, 35, 99
   strategic maneuver, 118, 119
   symmetrical wars, 31
   Vietnam War, 2, 73, 99–100
   War of Attrition (1969–70), 59
   war matrix, 9
   World War II, 2, 38, 41, 42, 99
   Yom Kippur War (1973), 2, 34–5, 51, 57, 100
military end state
   asymmetrical wars, 65
   classical Clausewitz doctrine, 51, 65
   Morocco crisis (1904–6), 28
   Operation Cast Lead (2008–9), 102
   Second Lebanon War (2006), 93, 94, 98, 99
   Sinai Campaign (1956), 61, 65
   US doctrine, 16, 17
   Yom Kippur War (1973), 3, 51
military strategy level, 2–3, 9
   asymmetrical wars, 65, 66
   guerilla paradigm, 110–15
   Lebanon War (2006), 89, 90, 91, 92
   Operation Cast Lead (2008–9), 104
   Sinai Campaign (1956), 62–3, 65
   Six Day War (1967), 33, 35
   Vietnam War, 73, 74
   World War II, 39, 41, 42, 119
   Yom Kippur War (1973), 3, 35, 44–5, 48, 50, 51, 52, 135$n$
military-operational centers of gravity *see* operational centers of gravity
Mitchell, Billy, 30
Moltke, Helmuth von, 19
Morocco, 27–9
Mutually Assured Destruction (MAD), 136$n$

Napoleon, 70, 78, 80
Napoleonic Wars, 51, 70, 80, 116

# INDEX

Nasrallah, Hassan, 86, 125
Nasser, Gamal Abdel, 43, 61, 62
NATO, Kosovo War, 23–4, 117–18
Naveh, Shimon *see* Systemic Operational Design
Nelson, Horatio, 80, 116
Netherlands
 non-state opponents, 70
 World War II, 38, 117

Operation Cast Lead (2008–9), 1, 102–8
 centers of gravity, 102, 104, 107
 Hamas, 102–4, 105, 106–8
 international community, 104–5, 106
 Israel Defense Forces (IDF), 102, 105, 107
 maneuver (ground mobile warfare), 104, 105–6
 operational level, 104, 105–6
operational art, 19–20, 25–6, 135$n$; *see also* Systemic Operational Design (SOD)
operational centers of gravity, 4, 7
 analytical tool, as, 7
 asymmetrical wars, 66
 classical Clausewitz doctrine, 1, 7, 11, 15–16, 25–7
 defined, 83
 Dien Bien Phu campaign (1953–54), 74
 First Lebanon War (1982), 75, 76
 guerilla paradigm, 113–14, 115, 117, 122
 non-state opponents, 71
 Six Day War (1967), 31
 symmetrical wars, 31
 US doctrine, 16
 World War II, 42
 Yom Kippur War (1973), 35
operational level, 2–3, 9, 83, 100
 asymmetrical wars, 65, 66
 guerilla paradigm, 110–15
 Operation Cast Lead (2008–9), 104, 105–6
 parallel warfare, 82
 Second Lebanon War (2006), 88, 89, 90, 92, 95
 Sinai Campaign (1956), 63–4, 65
 Six Day War (1967), 35
 strategic surrender, 83
 Systemic Operational Design (SOD), 19

US doctrine, 16
Vietnam War, 72–3
War of Attrition (1969–70), 59, 60
World War II, 38–9, 41, 42, 119
Yom Kippur War (1973), 3, 33, 34, 35, 43, 44, 45–6, 47, 49, 50–3, 54, 135$n$

Palestine Liberation Organisation (PLO)
 expulsion from Jordan, 75
 First Lebanon War (1982), 75, 76, 87–8, 95, 96, 135$n$
parallel warfare, 3–4, 77–84
 guerilla paradigm, 114–15
 Second Lebanon War (2006), 85–102
 Second Punic War, 4, 7, 80–2, 99, 116, 118, 136–7$n$
 World War I, 79–80
Patton, George, 78, 124
Philadelphi axis, 103, 104, 105
physical centers of gravity, 7
 guerilla paradigm, 76, 114, 115, 116
 Second Lebanon War (2006), 91, 94, 96
 Sinai Campaign (1956), 63
 US doctrine, 16
 *see also* abstract centers of gravity; centers of gravity; military centers of gravity
PLO *see* Palestine Liberation Organisation (PLO)
Poland, 61
political end state, 1–2, 130
 asymmetrical wars, 3, 65
 classical doctrine, 3, 65
 and international community, 11
 Morocco crisis (1904–6), 28
 Operation Cast Lead (2008–9), 102
 Second Lebanon War (2006), 3, 92, 97–8, 99
 Sinai Campaign (1956), 61, 62, 65
 World War II, 51
 Yom Kippur War (1973), 2, 3, 51, 100, 112
Prussia
 civilian–political stamina, 18, 123
 military decision, 18
 military doctrine, 18–19, 123, 136–7$n$
 multi-front wars, 18–19
 resources, 18
Punic War, Second, 4, 7, 80–2, 99, 116, 118, 136–7$n$

**143**

# INDEX

al-Qaeda, 70

Rafah, 32, 63
Reagan, Ronald, 135$n$
residual defiance capability
  Egyptian military, 55
  guerilla paradigm, 6, 76, 114, 115, 128, 129
  Hizbollah, 6, 90, 94, 114
  Vietcong, 73, 74
resources
  asymmetrical wars, 65
  Israeli wars, 5, 18, 23, 24
  outcome of the war determination, 2
  political will, 57–8
  Prussian/German doctrine, 18
  Second Punic War, 82
  Six Day War (1967), 32
  US doctrine, 5, 20, 21, 22–3
  Vietnam War, 2, 58, 73, 99–100
  war matrix, 9
  wars of attrition, 5
  World War II, 2, 38, 40–2, 57, 99, 134$n$
Revolution in Military Affairs (RMA), 12, 20–1, 123
  assimilation by IDF, 90
  critical/vulnerability nodes, 25–6
  defined, 132$n$
  exported US doctrine, as 125
  guerilla paradigm, 110–13
  Iraq wars, 23
Ribbentrop–Molotov Pact, 40
Rogers, William, 51
Rome *see* Punic War, Second
Rommel, Erwin, 124
Rundstedt, Gerd von, 26
Russia
  Chechnya campaign, 70, 87
  Kosovo War, 24
  Morocco crisis (1904–6), 27, 28, 29
  non-state opponents, 70
  *see also* Soviet Union

Sadat, Anwar, 43, 44, 50, 60, 125, 126
Saddam Hussein, 92
Scipio Africanus, 4, 7, 81–2, 99, 116, 118, 131
Serbia, 22, 23–4, 99, 117–18, 123–4
Shazly, Saad El, 43, 50
Sherman, William, 118, 136$n$
Short, Michael, 23–4
Sinai
  Six Day War (1967), 32
  Yom Kippur War (1973), 43, 44, 45, 52
Sinai Campaign (1956), 61–5
  "winning by not losing" principle, 62
Sinai Canal, War of Attrition (1969–70), 59, 60
"situation as a whole" approach, 4, 77, 78, 79, 125, 130
  Yom Kippur War (1973), 50, 51
Six Day War (1967), 31–3, 42, 99, 116
Soviet Union
  non-state opponents, 70
  Sinai Campaign (1956), 61–2, 65
  World War II, 39–40, 41, 117, 119
  *see also* Russia
Spain, Morocco crisis (1904–6), 27, 28
stamina *see* civilian–political stamina
Straits of Tiran, 19, 61
strategic centers of gravity
  defined, 20, 83
  First Lebanon War (1982), 75, 76
  guerilla paradigm, 115–21, 122
  maneuver (ground mobile warfare), 127
  non-state opponents, 71
  Revolution in Military Affairs (RMA), 20
  Second Lebanon War (2006), 76, 86–90, 91–2, 94–5, 96–7
  Second Punic War, 82, 118
  Six Day War (1967), 32, 116
  US doctrine, 16
  USAF strategic attack doctrine, 20–1
  War of Attrition (1969–70), 59–60
  World War II, 42, 117, 118
  Yom Kippur War (1973), 34, 35, 56–7, 118
Strategic Defense Initiative (SDI), 135–6$n$
strategic surrender, 82, 83, 121
Suez Canal
  Sinai Campaign (1956), 61, 63
  Yom Kippur War (1973), 47, 51
Summers, Harry, 73
Sun Tzu, 78, 125, 135$n$
symmetrical wars
  Six Day War (1967), 31–3, 42
  Yom Kippur War (1973), 33–6
Syria
  First Lebanon War (1982), 75
  Second Lebanon War (2006), 96
  war paradigm, 107–8

# INDEX

Yom Kippur War (1973), 33–4, 35, 42, 45, 118, 134*n*
Systemic Operational Design (SOD), 19–20, 25

tactical level, 2–3, 9
  Operation Cast Lead (2008–9), 102, 104
  Second Lebanon War (2006), 90, 95, 101
  Sinai Campaign (1956), 64–5
  Six Day War (1967), 33, 35
  Vietnam War, 72–3
  World War II, 39
  Yom Kippur War (1973), 34, 35, 44, 46–7, 49, 50, 51, 52–3, 54, 135*n*
Tal, Israel, 17, 25
techno-tactical level, 2
  Operation Cast Lead (2008–9), 102
  Second Lebanon War (2006), 90, 92, 101
  World War II, 39
  Yom Kippur War (1973), 47–8, 50, 51
Trenchard, Hugh, 20
Tu, Colonel, 73
Tunis, 75
Turkey, 62

UN Security Council Resolution 1701, 3, 106, 130
UN Security Council Resolution 1860, 105
United States
  avoidance of tests of stamina, 5, 21–2
  Civil War, 136*n*
  civilian and political support, 22
  defense dialogue with Israel, 23
  Iraq War (1991), 18, 23, 111, 116
  Iraq War (2003), 23, 70, 92, 111, 116, 118
  Kosovo War, 22, 23–4, 99, 117–18, 123–4
  military decision, 17, 18
  military doctrine, 5, 16–17, 18, 20–1, 22–5, 123–8
  Operation Cast Lead (2008–9), 105
  Sinai Campaign (1956), 61, 62, 65
  strategic attack doctrine, 20–1, 22–5
  Strategic Defense Initiative (SDI), 135–6*n*
  War of Attrition (1969–70), 60
  World War II, 39–41, 57, 124–5

Yom Kippur War (1973), 43, 44, 50, 52, 54, 56
  *see also* Revolution in Military Affairs (RMA); Vietnam War

Viet Minh, 70, 74
Vietcong, 70, 72–4, 86–7, 99, 117
Vietnam War, 54, 70, 71, 86–7
  civilian–political stamina, 2, 73, 99–100
  deterrence concept, 126
  military effectiveness, 2, 73, 99–100
  Operation Linebacker II, 58
  Operation Rolling Thunder, 24, 58
  resources, 2, 58, 73, 99–100
  Tet Offensive, 72–4
  US aerial attacks on Hanoi, 7–8, 116
  "winning by not losing" principle, 5
Von Naumann tables, 6

War of Attrition (1969–70), 43, 56, 59–60, 134–5*n*
weapons of mass destruction, 121–2
West
  military doctrines, 4, 78–9
  precision arms technology, 121
  techno-tactical superiority, 4
Westmoreland, William, 73
Westphalia, Peace of (1648), 70
"winning by not losing" principle, 5, 62, 86, 94, 110, 115, 125
World War I
  civilian–political stamina, 12
  naval campaigns, 79–80
  trench warfare, 79, 119
World War II
  asymmetrical war, as, 3, 38–42
  Belgium, 38–9, 54, 117
  Britain, 38–40, 57, 117, 124
  centers of gravity, 42, 116, 117, 118
  civilian–political stamina, 2, 12, 38, 40, 42
  France, 3, 38–9, 99, 116, 117
  Germany, 2, 3, 26, 38–40, 41–2, 54, 57, 70, 99, 116, 117, 119, 134*n*
  military effectiveness, 2, 38, 41, 42, 99
  Operation Overload, 26
  political end state, 51
  resources, 2, 38, 40–2, 57, 99, 134*n*
  Soviet Union, 39–40, 41, 117, 119
  strategic maneuver, 117, 118, 119
  United States, 39–41, 57, 124–5

INDEX

Yom Kippur War (1973)
  asymmetrical war, as, 33, 42–58
  centers of gravity, 33, 34, 35, 55–7, 116, 118
  civilian–political stamina, 2, 45, 52, 56, 57, 100
  classical Clausewitz doctrine, 33, 34, 35, 53
  deterrence concept, 126
  grand strategy level, 3, 43–4, 48, 50, 51, 52, 54–5
  international community, 2, 42, 45, 51, 100
  Israel Defense Forces (IDF), 33–4, 43, 45, 46–8, 50, 52, 53, 54–5, 56, 112, 135$n$
  military decision, 34–5, 50, 51, 52, 53–5, 112

military effectiveness, 2, 34–5, 51, 57, 100
military end state, 3, 51
military strategy level, 3, 35, 44–5, 48, 50, 51, 52, 135$n$
operational level, 3, 33, 34, 35, 43, 44, 45–6, 47, 49, 50–3, 54, 135$n$
political end state, 2, 3, 51, 100, 112
"situation as a whole" approach, 50, 51
symmetrical war, as, 33–6
Syrian involvement, 33–4, 35, 42, 45, 118, 134$n$
tactical level, 34, 35, 44, 46–7, 49, 50, 51, 52–3, 54, 135$n$
techno-tactical level, 47–8, 50, 51
theater of operations, 34, 55, 56, 121
"winning by not losing" principle, 5

www.ingramcontent.com/pod-product-compliance
Lightning Source LLC
Chambersburg PA
CBHW071412300426
44114CB00016B/2270